T0270705

Pests and Diseases in Fruit Crops

Dr S. Parthasarathy
Assistant Professor (Plant Pathology)
Amrita School of Agricultural Sciences,
Amrita Vishwa Vidyapeetham, Coimbatore

Dr P. Lakshmidevi
Teaching Assistant (Plant Pathology)
Institute of Agriculture, Kumulur, Trichy
Tamil Nadu Agricultural University, Coimbatore

Dr P. Yashodha
Associate Professor (Entomology)
Agricultural College and Research Institute, Thiruvannamalai
Tamil Nadu Agricultural University, Coimbatore

Dr C. Gopalakrishnan
Professor (Plant Pathology)
Department of Plant Pathology
Centre for Plant Protection Studies
Tamil Nadu Agricultural University, Coimbatore

CRC Press
Taylor & Francis Group
Boca Raton London New York

CRC Press is an imprint of the
Taylor & Francis Group, an **informa** business

Elite Publishing House

First published 2025
by CRC Press
4 Park Square, Milton Park, Abingdon, Oxon, OX14 4RN

and by CRC Press
2385 NW Executive Center Drive, Suite 320, Boca Raton FL 33431

CRC Press is an imprint of Informa UK Limited

British Library Cataloguing-in-Publication Data
A catalogue record for this book is available from the British Library

Print edition not for sale in South Asia (India, Sri Lanka, Nepal, Bangladesh, Pakistan or Bhutan).

ISBN13: 9781032823669 (hbk)
ISBN13: 9781032823676 (pbk)
ISBN13: 9781003504146 (ebk)

DOI: 10.4324/9781003504146

Typeset in Adobe Caslon Pro
by Elite Publishing House, Delhi

–EPH–

Contents

Preface

Management of pests and diseases are greatest importance for successful and profitable cultivation of fruit crops. Under these environments, important and relevant information were collected and compiled in a textbook form titled **Pests and Diseases in Fruit Crops**. This book is mainly intended for the Plant Protection courses of graduate students in the field of Agriculture, Horticulture, Botany, Forestry and Zoology. It is clear that young students are suffering from cultural shocks to shift from their environment. Semester system of education of B.Sc./M.Sc. (Ag.), B.Sc./M.Sc. (Horti.), B.Sc./M.Sc. (Forestry), B.Sc./M.Sc. (Botany), and B. Tech. (Horti.), students are quite dynamic for which the students are to be helped for changeover. We can identify their difficulties for comprehension of language, non-availability of textbooks for their semester system. There is a need to use simple language. The present book titled Diagnostics and Management of Pest and Diseases in Fruit Crops suite to the need of students. This book is written in simple understandable language dealing with various subject matters of Pests and Diseases. This book has been prepared with a specific purpose of importing complete comprehensive information about major diseases of horticultural crops in India and we hope that the students and readers will find this with much utility. This book aims to project the significant pest, disease, and nematode issues affecting essential fruit crops in the country. The management of pests in vegetable crops has been extensively addressed through many methods, including regulatory, physical, cultural, chemical, and biological approaches, host resistance, and integrated pest management strategies.

This book will function as a comprehensive manual for individuals involved in identifying and mitigating pest-related issues in fruit crops inside agricultural fields. Additionally, it can function as a pragmatic manual for farmers who cultivate vegetable crops. Moreover, this resource is a valuable reference point for policymakers, researchers, extension workers, and students. The book possesses potential utility in instructing both undergraduate and postgraduate curricula about crop protection. I thank all the scientists / publishers from which references were collected on various aspects of pests and disease aspects. I am sure that this book will serve as valuable text cum reader friendly textbook to the graduate and post-graduate students of agricultural universities.

Authors

About the Authors

Dr S. Parthasarathy is working as an Assistant Professor (Plant Pathology), in the Department of Plant Pathology at Amrita School of Agricultural Sciences. He has completed his PG and Ph.D. from Tamil Nadu Agricultural University, Coimbatore. He is a recipient of innovative fellowships such as the BIRAC Post Master's Innovation Fellow in DBT during 2015–17, the Canadian International Development Research Centre-Nano Project JRF during 2013–14, the UGC Non-Special Assistance Programme during 2012, and the Ministry of Human Resource Development Fellowship for UG during 2009, and he also received medals during his UG program and several meritorious awards in international and national conferences for his research contributions. A budding expert in Plant Pathology, he has published several research articles in refereed journals, authored 16 books, and contributed about 45 chapters to different books with national and international reputations. He also served as a member of the editorial boards of national journals and several international scientific societies. His areas of specialization are Biological Control and Molecular Plant Pathology.

Dr P. Lakshmidevi is a Teaching Assistant (Plant Pathology) in the Department of Plant Pathology, Institute of Agriculture. She has completed her Ph.D. from Tamil Nadu Agricultural University, Coimbatore. She is a recipient of innovative fellowships, a Fellow in RGNF from 2011-14, a Young Plant Pathologist 2019, and the Best Book Contributor Award 2022. She has conducted 2 Farm Testing trials, 4 Front Line Demonstrations, two farmers' field schools, and 43 on and off-campus training. She has published 4 books with ISBNs, 12 research articles, 29 conference papers, and 23 book chapters with ISBNs.

Dr P. Yashodha is an Associate Professor (Entomology), at the Department of Entomology, Agricultural College and Research Institute, Thiruvannamalai. She has completed 14 years of service and handled numerous undergraduate and postgraduate courses with a specialization in Biological controls, Insecticide Toxicology, Storage Entomology, and Molecular Ecology. She has guided many M.Sc. and over 5 Ph.D. students in Tamil Nadu Agricultural University. Numerous academic organizations have awarded her for her outstanding contributions to Agricultural Entomology. She has contributed to releasing crop varieties and technologies in Tamil Nadu. She has operated numerous funded projects in the university, including the DST and NAIP projects. She has published over 3 books with ISBNs, 40 research articles, 25 conference papers, 18 book chapters with ISBNs.

Dr C. Gopalakrishnan is working as a Professor of Plant Pathology at Tamil Nadu Agricultural University, Coimbatore. He has completed 28 years of service and has handled more than 50 courses at the undergraduate level and 34 courses at the postgraduate level. He has guided 3 Ph.D. and 7 PG students. Five Ph.D. students are currently working under his guidance in various areas of Plant Pathology. He has authored 6 books, 13 book chapters and published more than 50 research papers in national and international journals. He received a NATP-Team of Excellence fellowship in biological control during 2002-03. He has received one state-level and four university-level awards for his contributions to agricultural research, extension and teaching. He has attended over 30 national and international conferences, seminars and symposia. He was involved in releasing 8 crop varieties and 3 technologies to benefit the farming community.

1 Introduction

The global food system is undergoing a significant transformation due to changes in consumption patterns, namely, a shift away from rice, wheat, and pulses. Increased consumer awareness of health issues is what is driving this shift. Consequently, there is a notable shift in concentration toward horticulture crops. The growing awareness of the nutritional benefits of horticulture products in the human diet has increased the demand for fruits on the market. The advancements in production technology for horticultural crops and the recognition of these advancements as a method of diversification have played a substantial role in satisfying the growing demand for these products. Therefore, it is worth noting that India, although occupying only 2.5% of the world's land mass, is home to 16% of the global population. India's fruit production is recorded at 88.98 million metric tons, cultivated in an area of 7.21 million hectares, making it the second-largest fruit producer globally, following China. Not only does it rank highly in terms of total fruit output, but it also possesses the advantage of cultivating a diverse range of fruits. Among the prominent varieties are mango, banana, citrus, pomegranate, guava, grape, pineapple, papaya, apple, sapota, and litchi. Fruit cultivation is widespread across the country, with the state of Maharashtra making a notable contribution of 15.12 % to the overall fruit production in India. Following this are Andhra Pradesh (11.81 %), Gujarat (8.99 %), Tamil Nadu (8.28 %), Uttar Pradesh (7.74 %), Karnataka (7.47 %), Madhya Pradesh (6.40 %), Telangana (4.99 %), and Bihar (4.51 %). Mango, banana, grapes, and pomegranate are three significant fruit crops shipped to several nations, adhering to rigorous standards.

Fruit crops are susceptible to various pests that can cause damage to the trees, flowers, leaves, and fruits, resulting in reduced yields and fruit quality. Here are some common pests of fruit crops:

1. **Fruit Flies (*Drosophila* spp.):** These small flies lay eggs in ripe or damaged fruits,

leading to fruit infestations. The larvae (maggots) feed on the fruit, causing it to rot and become unsuitable for consumption.

2. **Codling Moth (*Cydia pomonella*):** A major pest of apples and pears, the codling moth larvae tunnel into the fruit, leaving brown frass and causing internal damage.

3. **Aphids:** These tiny, sap-sucking insects can infest various fruit trees and cause curling leaves, stunted growth, and transmission of viral diseases.

4. **Scale Insects:** Scale insects attach themselves to the stems and leaves of fruit trees, feeding on plant sap and weakening the tree. Honeydew excreted by scales can attract sooty mold and reduce fruit quality.

5. **Thrips:** Thrips damage fruit crops by feeding on the flowers and fruit, causing scars, deformities, and premature fruit drop.

6. **Mites:** Spider mites and rust mites are common pests that can damage fruit trees by sucking plant juices, leading to leaf discoloration, defoliation, and reduced fruit size.

7. **Leafhoppers:** Leafhoppers can transmit viral diseases to fruit crops and cause direct damage by feeding on plant sap.

8. **Whiteflies:** These small, flying insects feed on the undersides of leaves, causing yellowing, wilting, and reduced photosynthesis.

9. **Cherry Fruit Fly (*Rhagoletis* spp.):** A significant pest of cherries, the cherry fruit fly lays eggs in the fruit, leading to maggot-infested cherries.

10. **Oriental Fruit Moth (*Grapholita molesta*):** The larvae of this moth tunnel into peach and nectarine fruits, causing internal damage and increased susceptibility to diseases.

11. **Brown Marmorated Stink Bug (*Halyomorpha halys*):** An invasive pest, the stink bug feeds on various fruits, causing necrotic areas and rendering the fruits unmarketable.

12. **Leafrollers (Tortricid moths):** The larvae of leafroller moths roll themselves in leaves and feed on the foliage, causing defoliation and reduced photosynthesis.

Integrated pest management (IPM) practices, including cultural controls, biological controls, and judicious use of pesticides, are essential for managing pests effectively while minimizing environmental impacts and preserving beneficial organisms in the fruit crop ecosystem. Regular monitoring and early intervention are key to mitigating pest damage and ensuring a successful fruit harvest.

Fruit crops are susceptible to various diseases caused by fungi, bacteria, viruses, and other pathogens. These diseases can negatively impact fruit quality, yield, and overall tree health. Here are some common diseases of fruit crops:

1. **Apple Scab (*Venturia inaequalis*):** A fungal disease that affects apple trees, causing dark, scaly lesions on leaves and fruits, leading to defoliation and reduced fruit quality.

2. **Fire Blight (*Erwinia amylovora*):** A bacterial disease that affects pome fruit trees (such as apples and pears). It causes wilting, blackening of blossoms, and cankers on branches, leading to dieback and yield loss.

3. **Citrus Canker (*Xanthomonas citri* subsp. *citri*):** A bacterial disease affecting citrus trees, causing raised lesions on leaves, fruit, and stems, leading to defoliation and fruit drop.

4. **Powdery Mildew:** A fungal disease that affects various fruit crops, causing a powdery white growth on leaves, stems, and fruits, leading to reduced photosynthesis and fruit quality.

5. **Brown Rot (*Monilinia* spp.):** A fungal disease affecting stone fruits such as peaches, plums, and cherries. It causes brown rotting of fruits, leading to significant losses during storage and transit.

6. **Anthracnose:** A fungal disease that affects a wide range of fruit crops, causing dark, sunken lesions on fruits and leaves, leading to fruit rot and defoliation.

7. **Grapevine Downy Mildew (*Plasmopara viticola*):** A fungal disease affecting grapevines, causing yellowing and necrosis of leaves, and can lead to reduced grape quality and yield.

8. **Black Spot (*Diplocarpon rosae*):** A fungal disease affecting rose bushes, causing black spots on leaves, leading to defoliation and reduced plant vigor.

9. **Cucumber Mosaic Virus (CMV):** A viral disease affecting various fruit crops, causing mosaic patterns on leaves, stunted growth, and reduced fruit quality.

10. **Phytophthora Root Rot:** A soil-borne disease affecting various fruit trees, causing root rot, wilting, and eventual tree death.

11. **Verticillium Wilt (*Verticillium* spp.):** A fungal disease affecting a wide range of fruit trees, causing wilting, leaf chlorosis, and branch dieback.

12. **Apple Mosaic Virus (ApMV):** A viral disease affecting apple trees, causing mosaic patterns on leaves and reduced fruit quality.

Integrated disease management strategies, including the use of disease-resistant varieties, proper sanitation, cultural practices, and timely application of fungicides o other control measures, are essential for preventing and managing fruit crop disease effectively. Regular monitoring, early detection, and prompt intervention are vita for minimizing disease impact and ensuring healthy fruit production.

IPM aims to safely maintain economic, effective and long-term pest contro Generally, it contains suppressing pest populations to economic injury levels rathe than eradicating the pest and disease completely. Many pests negatively affec agricultural production in the world. Many methods are used by the producers t minimize the quality and quantity losses of these pests in agricultural production. Th main of these methods, which are considered for Plant Protection or Agricultur: Control, are cultural measures, quarantine measures, mechanical and physical method biological method, biotechnical method, chemical method and integrated pe: management, which expresses the combination of the necessary ones. Diagnosin and managing pests and diseases in fruit crops is crucial for ensuring healthy an productive fruit production. Fruit crops are susceptible to a wide range of pests an diseases, and effective management requires an integrated approach. Here's a step by-step guide to help with the process:

1. Regular Monitoring:

» Conduct regular field inspections and monitoring of fruit crops. Wal through the orchards, inspecting the trees and fruits for any signs of pest or diseases.

2. Identification:

» Accurately identify the specific pests and diseases affecting your fruit crop: Use field guides, online resources, or seek help from agricultural experts o local extension offices.

3. Cultural Practices:

» Implement cultural practices to create an environment less conducive t pest and disease development:

- Proper site selection and orchard design.

- Adequate spacing between trees for good air circulation.

- Pruning to remove diseased or infested branches and promote sunligh penetration.

- Proper irrigation and drainage to avoid water-related issues.

4. Use of Disease-Resistant Varieties:

 » Whenever possible, choose disease-resistant fruit tree varieties to reduce the likelihood of infections.

5. Biological Controls:

 » Introduce beneficial organisms that prey on or parasitize pests. Encourage natural predators like ladybugs, lacewings, and predatory mites to keep pest populations in check.

6. Traps and Monitoring:

 » Use traps and monitoring devices to detect and monitor pest populations. Sticky traps, pheromone traps, and visual surveys can help assess pest levels.

7. Pheromones:

 » Use pheromones to disrupt the mating and reproduction of certain insects, preventing population growth.

8. Chemical Controls:

 » If necessary, use chemical pesticides, but only as a last resort and in accordance with integrated pest management (IPM) principles.

 » Follow label instructions and use pesticides selectively to minimize negative impacts on beneficial organisms and the environment.

9. Disease and Pest-Free Plant Material:

 » Start with disease-free and pest-free planting material to prevent introducing problems into the orchard.

10. Record Keeping:

 » Maintain detailed records of pest and disease occurrences, control measures used, and their effectiveness. These records will help you track pest trends and make informed decisions in the future.

11. Education and Training:

 » Stay informed about the latest research and best practices for pest and disease management in fruit crops through workshops, seminars, and agricultural publications.

12. Timely Action:

> » Act quickly if pests or diseases are detected. Early intervention can preven
> further spread and minimize crop damage.

By following these steps and continuously monitoring your fruit crops, yo
can effectively diagnose and manage pests and diseases, ensuring a healthy an
bountiful harvest. Integrated pest and disease management is essential for long-tern
sustainability and minimizing the use of chemical inputs.

References

Alford, D. V. (2007). *Pests of fruit crops: a color handbook*. Elsevier.

Brunner, J. F. (1994). Integrated pest management in tree fruit crops. *Food Review International, 10*(2), 135-157.

Coates, L., & Johnson, G. (1997). Postharvest diseases of fruit and vegetables. *Plan pathogens and plant diseases*, 533-548.

Cooke, T., Persley, D., & House, S. (2009). *Diseases of fruit crops in Australia*. Csir publishing.

Drenth, A., & Guest, D. I. (2016). Fungal and oomycete diseases of tropical tre fruit crops. *Annual Review of Phytopathology, 54*, 373-395.

Fryer, J. C. F. (2008). *Insect pests of fruit crops*. Daya Books.

Hely, P. C., Pasfield, G., & Gellatley, J. G. (1982). *Insect pests of fruit and vegetable in NSW*. Inkata Press.

Kumar, J., Chaube, H. S., Singh, U. S., & Mukhopadhyay, A. N. (1992). *Plant disease of international importance. Volume III. Diseases of fruit crops*. Prentice Hall, Inc

Ploetz, R. C. (Ed.). (2003). *Diseases of tropical fruit crops*. Cabi Publishing.

Sharma, D. R., Arora, P. K., & Singh, S. (2017). Management of insect pests ir fruit crops other than citrus. *Theory and practice of integrated pest management Scientific Publishers Jodhpur (India)*, 410-425.

Srivastava, J. N., & Singh, A. K. (Eds.). (2022). *Diseases of Horticultural Crops Diagnosis and Management: Volume 1: Fruit Crops*. CRC Press.

Amla

INSECT PESTS

Shoot gall insect, *Rhodoneura emblicalis*, *Betousa stylophora*

Damage symptoms

- » Caterpillars cause galls on virtually every species.
- » A dark caterpillar that can be found inside galls on young branches.
- » In August and September, young caterpillars eat their way through the shoot until they reach the pith, at which point galls form.
- » Each gall contains a tiny opening through which the caterpillar's sticky, reddish-orange waste is expelled.
- » In severe cases, growth of the damaged branch may be stifled.
- » The galls have a length of 2.5–3.5 cm and a width of 1.5–2.0 cm.

Favorable conditions

- » This is a good time for rainy weather.
- » New growth is targeted during the monsoon and attacked by attackers.

Management

- » It's best to avoid a situation where branches are crammed together.
- » Galls on branches can be eliminated by cutting and burning the affected branches.
- » Spraying Chlorpyriphos (0.05%) at the start of the season is recommended if this pest is a yearly problem.

Mealy bug, *Ferrisia virgata*

Damage symptoms

» The insect creates a white mass over the developing tip, and then suckers the sap.

» Female insects, both nymphs and adults, congregate on the undersides of leaves, at the tips of stems, and even on the fruit itself, where they feed on the plant's cell sap and weaken it.

» A bending and twisting of the infected young shoots and a yellowing of the leaves are the telltale signs of infection.

» There is an abnormal amount of honeydew discharge.

» Stricken by a serious infestation, branches lose their leaves and eventually dry out.

» The flowers wither and fall.

» Loss of fruit before its time.

Pest identification

» **Nymph:** The nymph is a delicate yellow to white colour.

» **Adult:** Females reach sexual maturity as apterous, elongated insects with white waxy secretions covering their bodies.

Favorable conditions

» Climates that are both warm and humid are ideal.

Management

» Regularly raking the soil under a tree in the winter can help eliminate egg masses.

» Keeping the tree healthy and vigorous through careful cultivation.

» At the first sign of infestation, cut out and throw away any afflicted components.

» The population of newly hatched nymphs can be reduced by treating the soil with Neem cake at a rate of 300-500 g/tree.

» Nymphs can't climb trees if you wrap the trunk with a sheet of alkathene (400 gauge).

» Taking infected branches and leaves by the twigs and burying them.

- » Pest populations can be reduced through spraying with Monocrotophos, Spinosad (0.25 ml/L), or Quinalphos (0.05%) in cases of severe infestation.
- » Green lacewings (*Chrysoperla carnea*) were released into the wild.
- » The ladybird beetle, or *Chilocorus* sp., and the stag beetle, or *Cryptolaemus montrouzieri*, are both effective predators.

Fruit borer, *Deudorix isocrates*

Damage symptoms

- » The stages of a larva's development into a fruit are quite distinct.
- » Tender fruits are a favourite food for young larvae.
- » The full-grown larvae feed on ripe fruits.
- » Soft fruits that have been infested will first turn brown, and then black.
- » The fruit becomes hollow on the inside because the larva of this insect eats the seeds while boring into the fruit.
- » Usually, the larval entry hole will be the first visible sign of damage to fruits.
- » The borer hole may be leaking frass. These fruits never fully ripen, and instead wither and fall to the ground.
- » The rotting of ripe fruits begins on one side and spreads until the entire fruit rots and falls off.

Pest identification

- » Bright white eggs are deposited singly.
- » This is the female violet brown butterfly.

Management

- » Faulty fruit is gathered and thrown away.
- » Take away the pomegranate, guava, sapota, and tamarind as alternate hosts.
- » After the amla fruits have reached the pea size, spray them with Spinosad 0.25 ml/L or Carbaryl (2g/L). Depending on how severe the attack is, you may need to spray again in two weeks.
- » Use oviposition-preventing repellents like Neem oil 3% or NSKE 5%.
- » The Trichogramma chilonis population was increased by inundating the area four times at a rate of 2.5 lakhs/ha.

Bark borer, *Indarbela tetraonis*

Damage symptoms

>> Cut a path through the tree›s main trunk and its major branches.

>> The larvae create a sloppy web of silken threads.

>> Depletion of strength and vigour.

>> Falling production.

Pest identification

>> **Egg:** Eggs are round and are deposited singly in crevices in tre bark.

>> **Larva:** Larvae are 4.5-5.0 cm in length before they pupate; they range i colour from brown to black and have a shiny appearance with minimal hai

>> **Adult:** The adult moth has brown spots on its creamy white forewings.

Management

>> Maintain a tidy orchard.

>> Remove any peeling or broken bark and dispose of it.

>> An iron spike or wire inserted into the hole will quickly dispatch the larva

>> Spot treatments using 1 litre of water and 10 ml of Monocrotophos.

>> If the infestation is severe, take out the webs, inject a water emulsion c Chlorpyriphos (0.05%), or insert a swab of cotton wool soaked in 0.025 Dichlorvos, and then close the holes.

>> The entomopathogenic fungus *Beauveria bassiana* parasitizes the larvae i nature.

>> It shows promise as a biocontrol substance.

Fruit piercing moth, *Othreis materna, O. fullonica, O. ancilla*

Damage symptoms

>> Adult and larvae moths feed on the fruit's juice by puncturing it.

Management

>> Weeds like *Tinospora cardifolia* and *Cocculus pendules* need to be uprooted

>> Destroying rotten and rotten-looking produce.

» Orchard smoking.

» Moths are gathered with hand nets in the evenings.

» Weeds and creepers are collected for their semi-looping stems.

» Make use of light traps, and eliminate moths by maintaining a container of kerosene-soaked water below each trap.

» Carbaryl at a concentration of 2 g/L, or poison baits.

Aphid, *Setaphis bougainvilleae*

Damage symptoms

» Infested leaves become dry and yellow from a sugary discharge (honey dew).

» Growing tips of infected stems and leaves seem twisted and bowed.

» Aphid infestations are often accompanied by the presence of ants.

» Growth nodes on the young plants are infested.

» Both nymphs and adults stay on the undersides of leaves to extract the plant's sap.

» Heavy assault reduces the tree's vitality and growth, which in turn reduces the tree's ability to produce flowers and fruit.

Favorable conditions

» This pest is most active from July through October, with September being the highest month for occurrence.

Management

» Taking drastic action, such as snipping off and throwing away the afflicted leaf or stalk.

» Recognize and eliminate potential weed vectors.

» Set up yellow pan traps at the rate of four to five traps per acre.

» Apply a 3% solution of Neem oil in spray form.

» Apply a spray of either dimethoate (0.06%) or spinosad (0.25 ml/L).

» Aphid populations can be quickly reduced through the introduction of syrphids into the field.

» Green lacewings (*Chrysoperla carnea*) were released into the wild.

» Aphidius colemani is an example of a parasitoid.

» Predators include predatory mantids, green lace wings, ladybird beetles, and the coccinellids *Dicyphus hesperus, Scymnus, Cheilomenes sexmaculatus,* and *Chrysoperla zastrowi sillemi.*

DISEASES

Rust, *Ravenelia emblicae*

Symptoms

» Occasionally solitary or grouped, pinkish-brown pustules grow on leaves.

» Black pustules emerge on fruit at first, and then they grow into a ring.

» The pustules on fruit tend to cluster together and cover a wide region.

» By tearing open a papery casing, black spores are released.

» The appearance of fruit is unclean.

» Diseased fruits and leaves are brought on by teliospores from the invasive species *Ravenelia emblicae.*

Favorable conditions

» After September's monsoon, the weather is ideal.

Management

» Effectiveness of 0.5% Wettable Sulfur sprayed three times at monthly intervals beginning in July against rust disease.

» In order to prevent up to 95% of the disease, three sprays of 0.1% Bayleton were applied 15 days apart.

» A spray containing 0.2% Zineb is also useful.

» It's believed that the Banarasi and Chakaiya cultivars are relatively disease-free.

Anthracnose, *Colletotrichum* sp.

Symptoms

» Tiny brown or grey spots with a yellow edge develop first on leaflets, and then larger ones.

» The greyish core of the spots retains their spherical fruiting bodies.

» Acervuli, initially white dots that turn dark at the centre of fruits, are commonly organised in concentric rings.

» As a result, the fruit shrivels and spoils.

Management

» Pest control in the orchard.

» Throw out any rotten produce you find in the orchard.

» Apply Carbendazim spray (0.1% concentration) just before picking fruit.

» Apply a spray of Difolatan at 1500 ppm, Dithane M-45 at 100 ppm, or Carbendazim at 100 ppm immediately following harvest.

Soft rot, *Phomopsis phyllanthi*

Symptoms

» After only 2–3 days of infection, fruits show the telltale signs of round, smoke-brown to black blemishes.

» Eventually, the sick portions become discoloured in shades of olive brown, and water-soaked patches spread out from the centre to the fruit's tips, creating a "eye" shape.

» Dark brown spots develop on infected fruit, and the underlying tissues shrink and wrinkle.

» The fruit itself is misshapen.

» Fungi infect both immature and fully ripened fruit, but the latter are more likely to be eaten.

Favorable conditions

» Hot, damp weather is ideal for the spread of disease.

» Fungal growth is most successful between 29 and 32 °C.

Management

» Preserve fruit from harm.

» In the month of November, fruit is treated with Difolatan (0.15%), Dithane M-45 (0.1%), or Bavistin (0.1%).

Blue mold rot, *Penicillium citrinum*

Symptoms

- » The fruit becomes waterlogged and spotted with brown spots as a result.
- » Bright yellow, then purple-brown, and lastly bluish green appear in rapid succession as the condition advances.
- » Drops of a yellowish liquid are oozing out of the apple.
- » Fruits have a pungent odour.
- » The fruit takes on a beaded, feathery appearance in shades of bluish green.

Management

- » Safe fruit handling is essential. Blue mould can easily infect amla fruits if the skin is damaged in any way during harvesting or storage.
- » It's important to harvest without damaging the fruit in any way.
- » Storage areas should be kept clean.
- » Bleach or sodium chloride (1%), applied to fruit, prevents blue mould.
- » After harvest, treat with 0.1% Carbendazim or thiophanate methyl.
- » Mentha oil prevents decay of fruits when applied topically.

References

Afsah, A. F. E. (2015). Survey of insects & mite associated Cape gooseberry plants (Physalis peruviana L.) and impact of some selected safe materials against the main pests. *Annals of Agricultural Sciences*, *60*(1), 183-191.

Brennan, R. M. (2008). Currants and gooseberries. In *Temperate fruit crop breeding: Germplasm to genomics* (pp. 177-196). Dordrecht: Springer Netherlands.

Chaves-Gómez, J. L., Chávez-Arias, C. C., Prado, A. M. C., Gómez-Caro, S., & Restrepo-Díaz, H. (2021). Mixtures of biological control agents and organic additives improve physiological behavior in cape gooseberry plants under vascular wilt disease. *Plants*, *10*(10), 2059.

Haldhar, S. M., Agarwal, V. K., & Mani, M. (2022). Pests and Their Management in Indian Gooseberry/Amla. *Trends in Horticultural Entomology*, 817-831.

Johnson, L. P. (2010). A Survey of Production and Pest Management Strategies Used For Gooseberry Production Throughout Three Regions of the United States.

Jones, A. T., McGavin, W. J., Geering, A. D. W., & Lockhart, B. E. L. (2001). A new badnavirus in Ribes species, its detection by PCR, and its close association with gooseberry vein banding disease. *Plant Disease, 85*(4), 417-422.

Liefting, L. W., Ward, L. I., Shiller, J. B., & Clover, G. R. G. (2008). A new 'Candidatus Liberibacter'species in Solanum betaceum (tamarillo) and Physalis peruviana (cape gooseberry) in New Zealand. *Plant Disease, 92*(11), 1588-1588.

Mitchell, C., Brennan, R. M., Cross, J. V., & Johnson, S. N. (2011). Arthropod pests of currant and gooseberry crops in the UK: their biology, management and future prospects. *Agricultural and forest Entomology, 13*(3), 221-237.

Osorio-Guarín, J. A., Enciso-Rodríguez, F. E., González, C., Fernández-Pozo, N., Mueller, L. A., & Barrero, L. S. (2016). Association analysis for disease resistance to Fusarium oxysporum in cape gooseberry (Physalis peruviana L). *BMC genomics, 17*, 1-16.

Sengupta, P., Sen, S., Mukherjee, K., & Acharya, K. (2020). Postharvest diseases of Indian gooseberry and their management: a review. *International Journal of Fruit Science, 20*(2), 178-190.

3   Apple

INSECT PESTS

Codling moth, *Cydia pomonella*

Damage symptoms

- » Females deposit eggs on fruit or leaves, and the resulting yellow-headed black-headed larvae immediately begin feeding on the fruit.
- » After about three weeks of feeding inside the apple, the larvae pupate and overwinter elsewhere.
- » The protein-rich seeds are the primary source of sustenance.
- » This bug can reduce the marketability of apple crops by 30-70%.

Pest identification

- » **Larvae** - pinkish-white caterpillars with black or mottled black heads.
- » **Adult** - little, grey moth with chocolate brown patterns on its forewings.

Management

- » Sanitation in the field.
- » Remove caterpillar overwintering sites by scraping loose bark from tree trunks.
- » Those infected fruits and cocoons should be gathered and disposed of.
- » The tree trunks must be banded with corrugated cardboard.
- » Put out pheromone traps for sexual arousal.
- » Males en masse with codling moth lure traps.

» At the first sign of the moth, release egg parasitoid, *Trichogramma embryophagum*, at a rate of 2,000 adults per tree, with successive releases occurring once a week.

» Granulosis virus is produced commercially and sprayed at a rate of 8 x 108 virus capsids/ml every two weeks.

» Codling moth control was most effective when pheromone traps (E, E-10, 12-dodecadien-1-01) were used in conjunction with the introduction of the egg parasitoid Trichogramma embryophagum (2000 adults/tree, weekly intervals, 3 sprays).

San Jose scale, *Quadraspidiotus perniciosus*

Damage symptoms

» Infected areas of bark turn a bright pink colour.

» Fruits that have turned purple

» Young trees and branches lose vitality and die if they become infested.

» Infected branches take longer to burst their buds each spring.

» The malformed fruit loses its aesthetic appeal and commercial appeal.

» Scale insect infestations, particularly on young trees, can cause significant decline in tree health and even the loss of individual branches.

Pest identification

» Female: a little, dark yellow, oval, pustule with a sharp tip at one end.

» Males: elongated (oval) and have pale yellow bodies, dark eyes, and a lengthy protrusion from the base of the abdomen. Linear.

Management

» Grown without the presence of scales, these plants were hand-picked for the nursery.

» Use HCN gas or methyl bromide to kill pests in your nursery.

» Infested branches should be cut down and burned.

» Apply Phosalone 50 EC 0.05% or Fenitrothion 50 EC 0.05% in the summer to keep pests at bay.

» Diesel oil emulsion spray at 8-12 litres per tree during the winter (diesel oil 4.5 L, soap 1 kg, water 54 -72 L).

» Aphelinid endoparasitoid, *Encarsia perniciosi*, released at a rate of 2,000 individuals per afflicted tree once per spring.

» Parasitoids, such as *Prospaltella perniciosi* and *Aspidiotophagus sp.*, should b encouraged to do their thing.

» Predator coccinellids (*Chilocorus circumdatus/C. bijugus*) are released into th wild once every spring at a rate of 20 adults or 50 newborn grubs per tree

Woolly aphid, *Eriosoma lanigerum*

Damage symptoms

» Both nymphs and adults feed on the fruit and bark to get their nutrients

» Their eating causes twigs or roots to become knotted or galvanised.

» Water shoots are a more obvious target for galls than tree wounds.

» Galls on the roots are another symptom caused by colonisation below groun

» Plants, especially the young ones, are weakening and dying.

» Worm-eaten branches wither and perish.

» Woolly, white areas on the trunk.

» Honeydew, a sticky substance secreted by woolly aphids, falls down ont fruits and foliage.

» Honeydew can lower fruit quality by encouraging the growth of a fungu that leaves black sooty patches on the fruit and smells musty.

Pest identification

» Cottony white mats cover the underside of these purple aphids.

Management

» Take out the water shoots/suckers.

» Use a spray containing either 0.06% dimethoate or 0.025% methyl demeto e.c.

» When an infestation is discovered, release 1,000 mature Aphelinus ma parasitoids per tree.

» *Chilomenus bijugus, Coccinella septumpunctata, Hippodamia variegata, Syrphi confrator,* and *Chrysoperla* sp. are all predators.

» Make use of M 778, M 779, MM 14, MM 110, and MM 112, all of whicl are resistant root stocks.

Stem borer, *Apriona cinerea*

Damage symptoms

- » Insect larvae eat heartwood and burrow into the wood of a tree's limbs, stem, and trunk just beneath the bark.
- » The larvae bore 8-9 circular holes in the bark of each branch, spaced 10-15 cm apart, through which they exude frass.
- » Several holes have been cut into the trunk's interior.
- » A moderate wind gust can easily topple one of these trees.
- » Adults consume bark and leaves and girdle tender new growth, potentially causing its death.
- » There is a noticeable stunting effect on infected trees.

Pest identification

- » Grubs – Grubs are 60 x 12 mm in size and have a creamy yellow body with a dark brown tail and a white flat head.
- » Adults – Adults range in color from ash grey to greyish yellow, and have long horns that are topped with several prominent blackish tubercles towards the elytral base.

Management

- » Remove infested branches to prevent grubs from entering the tree itself.
- » Pruning and destroying recently attacked branches and killing older larvae by placing flexible wire into a live tunnel.
- » After plugging the holes with mud, insert a cotton wick dipped in gasoline, kerosene, 0.1% Dichlorvos, or 0.1% Methyl parathion (10 ml).
- » To get rid of the grub, inject 10 ml of Monocrotophos 36 WSC and cover the hole with damp clay.

Gypsy moth, *Lymantria obfuscata*

Damage symptoms

- » The larvae's widespread leaf-eating prevents fruit development.
- » Apple trees lose their leaves and vitality when infested with larvae, and in extreme cases, the infestation can even cause widespread destruction of apple trees.

» Caterpillars can fully defoliate their host plants during a severe attack stunting the trees' development.

» Without the host plant's leaves, no fruit will develop.

Pest identification

» Caterpillars range in size from 40-50 mm and are covered with tufts of hair

» Male moths are smaller and lighter in colour, while females are larger and more uniformly dark.

» Males are more active in the air and have browner bodies.

Spread

» Larvae in their first instar stay on the underside of leaves and are suspended by the large threads they make as the wind blows them from tree to tree.

Management

» The gathering and annihilation of oocyte clusters.

» Instead of spending money on insecticides, it was more cost-effective to use burlap bands 30 cm wide and staple them around the trunks of the trees that were being attacked. Larvae congregate during the daytime under burlaps, and this provides an opportunity to capture them.

» Spraying tree basins with grass mulch and spraying the tree trunk and foliage with concentrations of 0.04% Chlorpyrifos, 0.05% Endosulfan, and 0.0028% Deltamethrin resulted in the maximum larvae death.

» The density of larvae was reduced by 25-63% after field administration of strains of *L. obfuscata* multiple nucleopolyhedrovirus (LyobMNPV) at 2.51012 OBs/ha.

» A 500 g dose of disparlure [(Z)-7, 8-epoxy-2-methyloctadecane] was found to be efficient in capturing gypsy moth populations.

» Biological control using the parasitoids *Anastatuis kashmiriensis* (an egg parasitoid), Cotesia melanoscela (a larval parasitoid), *Glyptapantelos indiensis* (a larval parasitoid), and *G. flevicoxis* (a pupal parasitoid).

Cottony cushion scale, *Icerya purchasi*

Damage symptoms

» Sap from the leaves, branches, and stem (phloem) to weaken its host.

» Leaves are turning yellow.

» Defoliation and the death of young branches can result from severe infestations of these pests.

» Apple tree yields can be drastically decreased due to high human densities.

» Produces honeydew, which often attracts ants and black sooty mould.

» Adults and juveniles alike feed on the sap of leaves and branches.

Pest identification

» Nymph - Pink-colored creeper with long antennae and a cluster of hairs

» Female with a fluffy ovisac

Management

» Pick rootstock that hasn't been infested by any disease or pests.

» The infected plant components must be gathered and disposed of.

» Neem oil (2% NSKE) spray application.

» Chlorpyriphos 20 EC 0.04% spray applied with adhesive.

» Chilocorus nigritus and other predatory coccinellids were released into the wild.

» Biological control with the vedalia beetle (Rodolia cardinalis) and the parasite fly (Cryptochaetum iceryae).

Tent caterpillar, *Malacosoma kashmirienses*

Damage symptoms

» When larvae feed on the leaves of their host plants in great numbers, they can cause severe defoliation and so harm.

» For protection, larvae construct huge silken tents over food sources, such as leaves.

» Complete defoliation of tiny trees by larvae may not kill them but certainly inhibits their development.

Pest identification

» The Styrofoam-like brown or grey eggs measure about 0.06 inches in length.

» The western tent caterpillar larva is hairy and a drab yellow-brown colour;

it is spotted all over with blue and orange in neat rows.

» The wings of an adult moth can range from pale to dark brown.

Management

» It is possible to hand-pick tent caterpillars from trees and place them in bucket of soapy water.

» To prevent future damage, it may be necessary to remove nests from tree

» Use insecticides such as Spinosad, Insect-Off, or Azadirachtin (Neem oi spray.

» Diazinon, Phosmet, Azinphos-methyl, and Chlorpyrifos are all pesticide that can be employed in dire emergencies.

» *Bacillus thuringiensis* var. kurstaki has the most effect on younger ter caterpillars.

» The caterpillars are parasitized by the larvae of a fly called a tachinid.

European red mite, *Panonychus ulmi*

Damage symptoms

» To inflict the distinctive leaf damage known as bronzing.

» In order to get at the chlorophyll inside a leaf, mites use their needle-lik mouth parts to puncture the cells.

» Stippled and even bronzed symptoms emerge on the leaves when th population is big enough.

» Defoliation may occur in the event of a particularly severe infestation.

» Their reaction to micronutrients and growth regulators is modified, whic in turn impacts leaf efficiency and productivity.

» Early in the season, heavy mite feeding can have a negative impact on tre development and yield, as well as on fruit bud formation and, consequentl yields the following year.

Pest identification

» Eggs: a dull red, slightly flattened (onion-shaped), and have a hair-like ster sticking out of the top.

» Nymphs: Nymphs (larvae) have three sets of legs when they first hatcl These older nymphs, like adults, have four legs each.

» Adults: The full-grown female measures about 0.40 mm in length and has a dark brown-red coloration with rows of spots that have raised "spines" on their backs. The male is noticeably more diminutive than the female (0.28 mm in length), paler or drabber in coloration (with a pointed abdomen), and has longer legs.

Favorable conditions

» During the months of May and June, the weather is typically hot and dry.

» Nitrogen content in leaves is very high.

» Making use of poisonous chemicals that would kill off the pests' natural predators.

Management

» The mite population can be controlled by using natural predators such the coccinellid *Stethorus punctum*, the anthocorid bug *Orius sp.*, the *Chrysoperla sp.*, and the *Typhlodromus pyri*.

» To prevent dusty conditions that pest mites thrive in, plant well-managed cover crops in the spaces between rows.

» For efficient control of European red mite, apply a delayed dormant oil (from tight cluster through pink) to suffocate the mites' eggs.

DISEASES

Scab, *Venturia inaequalis*

Symptoms

» Foliar lesions typically manifest as olivaceous patches on the underside of leaves, which eventually turn a dark brown or black and develop a velvety texture.

» The spots appear to radiate from the young leaf and have a feathery edge.

» Lesions on older leaves have sharper margins and a clearer outline.

» In some cases, the lesion might cause a convex surface to form with a concave portion on the other side.

» The blade of the leaf is bent, shrunken, and distorted when the disease is severe.

» Small, rough, black, round blemishes can be seen on the fruit.

» On ripe fruits, the lesions' centres turn corky, and the skin around them takes on a yellow halo.

Favorable conditions

» The rapid progression of the disease is facilitated by the presence of high relative humidity, high rainfall, and temperatures between 16 and 20 °C.

» The disease thrives in cool, damp environments, such as those caused by rain or snow at higher altitudes or in shady areas of the orchards.

Survival and spread

» The *V. inaequalis* fungus survives the winter in the decaying apple leaves i infected the previous year. Spores of scab can survive the winter in crevice in the bark or even in the buds of infected trees.

» Ascospores are the primary vector for transmission, while conidia disseminat through the air.

Management

» Maintaining a tidy farm by raking and burning dead leaves.

» Potassium bicarbonate in a formulated form, sold under the brand name Armicarb 100 and containing surfactant components, proved more efficien than plain potassium bicarbonate.

» Petal fall followed by short bursts of spraying with either 0.2% Captan o 0.25% Dodine, or a single application of 0.3% Difolatan at the green bud stage followed by 0.2% Captan at petal fall.

» After the petals have fallen, spray twice with either 0.2% Captan or 0.25% Mancozeb, spaced out by 10-15 days. This is in addition to the 2% Urea spray used in the fall before leaf fall and again just before bud break.

» *Microsphaeropsis ochracea*, a mycoparasite, shows promise in decreasing the initial ascospore inoculum by as much as 70-80%.

» There are several cultivars with single-gene resistance, including Liberty Modi, Topaz, Pristine, and Ariane.

Powdery mildew, *Podosphaera leucotricha*

Symptoms

» It causes problems for the fruits, flowers, and new growth.

» The fungus manifests itself on leaves, especially the undersides, as felt-like patches or a dense mat.

» Chlorotic spots or patches on the upper surface of a leaf may be the result of an infection on the leaf's underside.

» Infected leaves have a wrinkly, curled, or rolled up appearance along their margins.

» During the summer months, leaves often fall off early if an infection is particularly bad.

» Infected young plants typically do not thrive.

» The mildewed shoots will be defoliated after a severe infection, which will slow their growth and reduce the number of buds they produce.

» The fruits don't develop normally, showing signs of stunting and rosetting.

Favorable conditions

» A favourable environment for conidial germination includes a relative humidity of more than 90% and a temperature of 19° to 22°C.

» Climate scientists believe the dry weather helped spread the disease.

Survival and spread

» Mycelium forms over the winter in infected vegetative buds and fruits.

» Conidia carried by the wind play a secondary role in the disease's transmission.

Management

» Disease can be reduced through proper pruning techniques such as cutting back lateral branches, thinning out the main stem, and removing and destroying any infected tissue.

» Lime sulphur (1:60 dilution) applied beginning in the latent stage and continuing fortnightly through the bloom season effectively suppresses the disease.

» Before flowering and while in full bloom, a 0.3% Wettable Sulfur spray is advised.

» Binapacryl applied as a spray at 450 g active ingredient per hectare in 225 litres of water is highly effective.

» Application of 0.1% Carbendazim is also effective against powdery mildew

» Contrasting Apple and its competitors. It has been discovered that the varieties Maharaja Chunth and Golden Chinese are resistant to powdery mildew.

Crown, collar, and root rot, *Phytophthora* spp.

Symptoms

» The infection manifests initially around the neck and then moves downward. Bark at the soil level gets slimy and rots resulting in cankered areas

» Chlorotic, red-veined, and margined leaves are a telltale sign that a tree has been assaulted.

» When the pathogen attacks the lower trunk at the soil line, it is called crown rot, whereas root rot primarily affects the tree's root system.

» Collar rot manifests as a depressed canker in the bark of the lower half of the scion, and can be any shade of brown, grey, or purple.

» When the bark is removed off a tree's trunk, the normally green cambium is shown to be orange or brown.

» The secondary roots may peel away from the main root and be a dark brown or orange colour.

Favorable conditions

» Solangiospores are most likely to form in soils with high moisture and temperatures between 20 and 25 °C, while oospores are more likely to form in soils with low moisture at the same temperature.

Survival and spread

» Most orchard soils are suitable for the fungus, although ideal conditions include a soil temperature of 12°-20°C and a pH of 5-6.

» Oospores can also be spread through contact with infected bark or pieces of bark.

» Although zoospores are rarely the cause of disease in apples, they can infect the plant if they are released in the soil.

Management

» Don't oversaturate the soil with irrigation water; use the right amount.

- » Remove any debris and infection from the diseased area of the collar before applying the Bordeaux or Chaubatia paste.

- » Rootstock grafting takes place 30 cm above the ground.

- » Fungicides such as Dithane M-45 or Metalaxyl are sprayed liberally around the tree's base.

- » Soil solarization with the application of antagonists such as *Trichoderma viride, T. harzianum, Enterobacter aerogenes,* and *Bacillus subtilis.*

- » *Streptomyces lydicus* (Actinovate), *Trichoderma* (Bio-TAM), *Bacillus amyloliquefaciens* (Triathlon), and *Bacillus subtilis* (Baccontrol) are all effective biocontrol agents (Serenade).

- » A number of rootstocks have been identified as resistant: M2, M4, MM104, MM113, and MM114.

Blotch, *Marssonina coronaria*

Symptoms

- » Blotch symptoms typically manifest as dark brown blotches on the upper surface of leaves, with sizes ranging from 5 to 10 cm in diameter.

- » If the weather is just right, these specks will join together to form enormous dark brown blotches, while the surrounding areas will turn a sunny yellow.

- » This yellowing progresses toward the petiole and causes midseason defoliation.

- » Disease symptoms include irregular necrotic spots on fully developed leaves.

- » Apples that are getting close to maturity can often be seen dangling from the defoliated trees in impacted orchards.

- » Failure of the crop the following season is the outcome of repeated early leaf shedding.

Favorable conditions

- » The quantity of conidia dispersed in the apple orchard was positively and strongly linked with rainfall and relative humidity. Defoliation occurs when the temperature rises above 20 °C in June and July and the relative humidity rises above 70% and remains there for more than five or six days.

Survival

- » The decaying leaves provided a suitable environment for the survival of fungi.

Management

» Leaf raking and leaf burning are traditional autumn activities.

» Kresoxim- methyl (0.1% Ergon 44.3 SC) and Pyraclostrobin (0.1% Cabr Top 60 WG), both of the Strobilurin type, can be used to effectively mana; this disease in situations of high humidity and rainfall.

» At the walnut stage, spray with a mixture of Mancozeb 75 WP (500g) ar Carbendazim 50 WP (100g) per 200 L of water; at the fruit developme: stage, spray with Propineb 70 WP (600g) per 200 L of water; at the fru development stage, spray with Dodine 65 WP (150g) per 200 L of wate and spray with Carbendazim 50WP (100g) 20-25 days before harvest.

Canker, *Physalopora obtusa*

Symptoms

» Stem canker, leaf spot (frog eye leaf spot), and fruit rot (black rot) are tl three stages it passes through; the canker phase is the most damaging.

» Sunken, reddish brown lesions form on the trunk and branches, eventual becoming smoky and forming a series of alternate rings.

» Down below, the wood is a rich, crimson colour.

» Needles, spurs, and branches above the canker die.

» Over the bark of the affected twigs or at the canker's periphery, mar pimple-like protuberances may form.

Favorable conditions

» Conditions ideal for the spread of disease are high humidity (>75% humidit and temperatures between 20 and 22 °C.

Survival

» The pathogen overwinters as dormant mycelium and fruiting structure i cankered mummified fruits and on dead wood.

Management

» Keep yourself safe from any potential mechanical harm.

» Treat cuts by applying wound dressings.

» Extend the girdle and cut off the affected limbs.

» Bordeaux (Cu So$_4$ - 1 part, lime and linseed oil - 2 parts each) or Chaubati paste should be applied to cankers after they have been scraped clean with a sharp knife (Lead oxide – 1 part, Copper carbonate – 1 part, Linseed oil – 11 parts).

» Take out all the dead branches and clipped fruit, and set them on fire.

Soft rot, *Penicillium expansum*

Symptoms

» The fruit's stem end is the first to show signs of the young spots, a light brown watery rot.

» Wrinkled skin and more rotting occur as the fruit ripens.

» Odors of a distinct mustiness are emanating.

» A sprig-like growth with a bluish-green colour appears in damp environments.

Survival and spread

» Bug bites and other skin punctures sustained in storage or transit are major vectors of infection.

Management

» Fruits must be picked, graded, and packed with extreme caution to prevent damage to the produce.

» Any rotten or broken fruit must be thrown away.

» After being packed, fruits need to be kept refrigerated.

» Sodium metabisulfite fumigation with sulphur dioxide proved very successful in halting the spread of the disease.

» Putting paraffin oil and spicy mustard oil on an apple prevented it from spoiling.

» Fruits that were packaged in SOPP-impregnated wrappers were somewhat shielded against *P. expansum*.

» At 25 °C, benomyl at a concentration of 1000 parts per million was the most effective therapy for soft rot management (Roy, 1975).

» Fruits were effectively protected from the disease when they were dipped in either 1000 ppm Diphenylamine for 5 minutes or 500 ppm Aureofungin for 20 minutes.

Fire blight, *Erwinia amylovora*

Symptoms

- » Blossom, fruit spur, and new shoot blight are the first signs of the disease in spring.
- » The wilting begins at the tips of the terminal twigs and moves downward and outward to the branches.
- » When infected flowers are exposed to water, they quickly wilt, turn from pale to dark brown, and either fall off the tree or die there.
- » The pedicel may act as a pathway for bacteria to get from the fruit spur to the leaves.
- » They are now following the more opaque midrib and major veins.
- » Infected areas may ooze bacterial fluids.
- » The fungus can infect vegetables and fruits.

Favorable conditions

- » Rain and temperatures exceeding 24 °C.

Survival and spread

- » Infected twigs and branches are the bacteria's main source of sustenance.
- » Bacterial ooze is easily transmitted during wet periods due to the combined efforts of rain splash and insects.

Management

- » Abolish and destroy the offending components.
- » During the summer months, you should prune away any diseased branches or new root growth.
- » Streptomycin preparations should be applied twice. When daily temperatures are over 65°F and a wetting event is expected within 24 hours, the first spray can be applied at any time after the first flowers open. If disease conditions persist through late bloom, spray again every 5 to 7 days.
- » *Bacillus subtilis* QST713 (Serenade Max), *Bacillus pumilis* QST2808 (Blight Ban), and *Pseudomonas fluorescens* A506 (Blight Ban) for biocontrol (Sonata)

References

Gladieux, P., Zhang, X. G., Afoufa-Bastien, D., Valdebenito Sancolorza, R. M., Sbaghi, M., & Le Cam, B. (2008). On the origin and spread of the scab disease of apple: out of central Asia. PloS one, 3(1), e1455.

Grove, G. G., Eastwell, K. C., Jones, A. L., & Sutton, T. B. (2003). Diseases of apple. In Apples: botany, production and uses (pp. 459-488). Wallingford UK: CABI Publishing.

Harris, D. C. (1991). The Phytophthora diseases of apple. Journal of horticultural science, 66(5), 513-544.

Khan, A. A., Wani, A. R., Zaki, F. A., Nehru, R. K., & Pathania, S. S. (2018). Pests of apple. Pests and their management, 457-490.

Mazzola, M., & Manici, L. M. (2012). Apple replant disease: role of microbial ecology in cause and control. Annual Review of Phytopathology, 50, 45-65.

Oerke, E. C., Fröhling, P., & Steiner, U. (2011). Thermographic assessment of scab disease on apple leaves. Precision agriculture, 12, 699-715.

Paulson, G. S., Hull, L. A., & Biddinger, D. J. (2005). Effect of a plant growth regulator prohexadione-calcium on insect pests of apple and pear. Journal of economic entomology, 98(2), 423-431.

Shaw, B., Nagy, C., & Fountain, M. T. (2021). Organic control strategies for use in IPM of invertebrate pests in apple and pear orchards. Insects, 12(12), 1106.

Sherwani, A., Mukhtar, M., & Wani, A. A. (2016). Insect pests of apple and their management. Insect pest management of fruit crops. New Delhi: Biotech Books, 295-306.

Zhou, H., Yu, Y., Tan, X., Chen, A., & Feng, J. (2014). Biological control of insect pests in apple orchards in China. Biological Control, 68, 47-56.

4

Avocado

INSECT PESTS

Thrips, *Scirtothrips perseae*

Damage symptoms

- » Premature leaf drop can be caused by a thrips infestation, which can reach epidemic proportions on young avocado leaves from late fall to early spring.
- » However, larvae and adults of avocado thrips cause the most damage to crops by scarring immature fruit in the spring.
- » Depending on how bad the scarring is, the entire surface of the fruit might turn brown and take on a "alligator skin" appearance.
- » When avocado thrips feed on immature fruit, the resulting scars are long and narrow.

Favorable conditions

- » *S. perseae* prefers the cool temperatures that prevail from late winter through early summer, when avocados are at their most productive.

Spread

- » *S. perseae* may be carried great distances by strong winds.
- » Adult *S. perseae* are known to travel great distances quickly and are thought to be spread through the transportation of infested harvesting bins.

Management

- » For planting, only use certified pest-free plant material.

» Mulch spread around avocado trees can prevent thrips from developing into adults and reduces their numbers by as much as half when coarse composted organic yard is used.

» When sprayed with Sabadilla, Spinosad, or Abamectin, avocado thrips are effectively eradicated.

» The *Orizaba frankothrips* (a common predatory thrips in avocado orchards) Avocado thrips larvae can be used for orchard-wide pest control.

eed moth, *Stenoma catenifer*

Damage symptoms

» Galleries made by moths cause a decrease in branch vitality and flower production.

» The larvae of some moths are able to eat their way into any kind of fruit and completely ruin the seed and pulp.

» Easy-to-see symptoms in fruit include holes, the look of a white exudate pouring out, and the presence of frass.

» When a fruit is cut open, the space that once contained the seed may now be filled with frass.

» The pulp and the seed may get separated as the fruit ripens, the seeds may sustain damage from larval feeding, or the larvae may consume and kill the seeds entirely.

pread

» The moths spread regionally via flight and by moving around in fruit.

» The dispersal of fruits over greater distances is likely to occur due to their transportation.

Management

» Disposal and recycling of rotten or otherwise unusable fallen fruit.

» Parasitism of larvae by *Apanteles sp.* is reported to occur in Venezuela at a rate of 31% to 33%.

» The sex pheromone can be employed for capturing in export orchards since adult male *Stenoma* are particularly attracted to it.

» The parasitic wasps *Trichogramma pretiosum* and *Trichogrammatoidea annulata* were set free in large numbers.

Looper, *Sabulodes aegrotata*

Damage symptoms

» Dense vegetation, an abundance of fresh leaves, and cloudy, cool weath
 are all factors that might contribute to damage. The damage to the leav
 is most noticeable on the tips of the branches.

» Identifiable brown membranes are left behind by very young larvae that e
 exclusively on the leaf surface.

» In many cases, only the midrib and major veins of a leaf are left after old
 larvae have finished munching on it.

» Lone leaves can be devoured by fully formed larvae in about one day.

» Excessive feeding might cause sunburn, which in turn can lower yield tl
 following year.

» Caterpillars do significant economic harm by eating fruit.

» Fruit can be chewed by both young and old larvae.

» Misshaped fruit can occur if young fruit is consumed.

» Surface scarring caused by chewing is common and can lead to the fru
 being culled or devalued.

Management

» When trees are cut down to expose more sunlight, noneconomic loop
 populations thrive.

» If administered at high enough concentrations and with enough coverag
 Kryocide (cryolite) at 20 lb a.i. /acre and Thuricide *(Bacillus thuringiensi*
 at 4 lb a.i. /acre effectively inhibited omnivorous looper larvae.

» There was success in spraying *Bacillus thurungiensis* or Spinosad.

» The omnivorous looper population has been kept in check by the release
 Trichogramma platneri in late spring or early summer, during peak moth egg layin

» The most effective predators are wasps, specifically the *Trichogramma eg*
 parasitoid and the three larval parasitoids.

» Internal parasites of larvae include the solitary *Apanteles caberatae* and th
 gregarious *Meteorus tersus.*

» Each looper is parasitized by an externally feeding larva of the *Habrobraco*
 (=Bracon) xanthonotus.

Brown mite, *Oligonychus punicae*

Damage symptoms

» The avocado brown mite typically feeds along the veins of top leaves.

» Chlorophyll loss from mite feeding inhibits photosynthesis.

» Defoliation can occur if an infestation is severe enough.

» If there are 300 mites on a single leaf, the plant could die.

Favorable conditions

» Conditions such as high temperatures, the use of malathion, and the accumulation of dust or ash from fires can all contribute to an outbreak.

Management

» Narrow-range oil, Etoxazole (Zeal), and Abamectin are three examples of chemical pesticides (Agri-Mek S).

» The spider mite killer (*Stethorus picipes*), green lacewings (*Chrysoperla sp.*), predatory mites, etc. are all effective biological control agents.

DISEASES

Root rot, *Phytophthora cinnamomi*

Symptoms

» In the later stages of the disease, the leaves of afflicted trees may be a pale green colour, wilt, or perhaps fall off entirely, and the terminal branches will die.

» The rootlets decay, the trunk gets girdled, and the leaves fall off because of the disease, which causes the tree to die.

» The bark of infected trees turns brown and develops longitudinal fractures, and the gummy substance on their surface is the first symptom of the disease.

» The girdling effect causes the bark to deteriorate and the tree to dry out.

» Trees that have been damaged experience leaf wilting and eventual leaf fall.

» Extreme cases typically result in the death of trees due to the darkening and rotting of their feeder roots.

Favorable conditions

» Disease is more likely to appear in areas with poorly drained soil and heavy irrigation.

» Injury to the roots at the base of the stem and the accumulation of soil around the collar area also promote infection.

Survival and spread

» The root-infecting fungus *Phytophthora cinnamomi* is widespread in soil and can live there for years if conditions are unfavourable.

» The infection rarely spreads above ground and often affects only the roots and lower trunk.

» Zoospores, or motile asexual spores, play a crucial role in the rapid dissemination of the virus.

Management

» It is advised to get seedlings from areas free of known diseases.

» Soil drenching with 0.2% Aliette at the base of the plant is an efficient method of disease management.

» Aliette is also effective when injected into the trunk.

» The best chemical remedy for root rot is phosphonate fungicide.

» Spray the ground with Metalaxyl grains.

» Typical Avocado. It has been said that Duke can withstand the disease.

Anthracnose, *Colletotrichum gloeosporoides*

Symptoms

» Both the leaves and the fruit of the avocado plant are susceptible to anthracnose infection.

» However, once the fruit is infected with the anthracnose fungus, the quality of the ripening fruit quickly declines.

» The lesions on the fruit's skin start out small, light brown, and circular, but they eventually get larger, become slightly sunken in the middle, and darken to a brown or black tint.

» The black decay causes lesions to turn dark and sunken, and it can spread to infect a sizable amount of the fruit as well.

» Indicators of ripeness are often a precursor to the premature dropping of fruits.

Favorable conditions

» At about 28 °C, a disease is most likely to spread.

» From the time of fruit set to the time of harvest, fruits are vulnerable to infection during periods of prolonged rainfall and/or high relative humidity.

Survival and spread

» Dead avocado leaves and twigs in the plant canopy or on the ground provide the fungus with food between fruiting cycles.

» Infected or colonised avocado twigs (living or dead) or infected leaves can transfer the fungal conidia (spores) to avocado fruits via splashing generated by raindrop impact.

Management

» Vacant the field of any plant remnants that have fallen to the ground.

» To increase air circulation under an avocado tree's canopy, prune it.

» Clear away any debris or decaying leaves from the canopy.

» Infections can be avoided by spraying healthy tissue with a fungicide (such as copper or Azoxystrobin).

Leaf and fruit spot, *Pseudocercospora purpurea*

Crop losses

» For orchards that go untreated, this pest can cause yield reductions of up to 69%.

Symptoms

» Affects the leaf edges, causing them to become marked with tiny, angular dots that are brown at first and eventually purple. A lot of the spots have golden rings around them.

» As the spots grow and join together, they can cause widespread deformity in the leaves.

» Young trees may lose their leaves if the disease spreads up the stem.

» Small, scattered, somewhat sunken flecks appear on fruit, too, and like the previous ones, they have the potential to crack and let more diseases in.

» Fungi can occasionally cause flesh infections.

» Wet conditions cause the patches to get covered in a greyish mycelium and fungal spores.

Survival and spread

» Conidia, which form on sick organs in moist and warm environments, are the primary infectious agents.

» Conidia can be carried to the foci of infection by water, wind, or insects.

Management

» Destruction of diseased branches and plants.

» Copper oxychloride (0.2-0.3%) alone or in conjunction with systemic fungicides like Azoxystrobin (0.3%) was applied once each month from October through January.

» The use of pre-harvest sprays containing *Bacillus subtilis* (B246 isolate, 107 cells ml-1) alone or in combination with other fungicides.

» Edranol and Cvs. Hass are more resistant to the disease.

Powdery mildew, *Oidium* sp.

Symptoms

» Powdery mildew is characterised by dark green patches on immature leaves that are covered with a dry, powdery covering of the causative fungus.

» As the leaves age, the purple-brown dots develop a white fungal growth.

» After a while, the fungal covering on the underside of these spots wears off, leaving behind net-like brown blotches.

» A net-like lesion may develop into a yellow spot on the upper leaf surface.

Management

» It may be enough to just prune trees in a way that lets more light and air into the leaves.

» Sulfur fungicide sprayed on leaves may aid in preventing the spread of disease.

Scab, *Sphaceloma perseae*

Symptoms

» Leaves, stems, and fruits may be affected by lesions.

» Brown, corky, elevated, oval, or irregular shaped patches and lesions appear on the fruit's surface as a result of fungal infection.

» As the disease progresses, the spots may merge, giving the entire fruit a rough, russet appearance.

» Little dark brown dots appear on leaf veins.

» Leaves may wilt and twist if they are severely infected.

» It is possible for leaf veins, pedicels, and twigs to develop scaly, massive sores.

Favorable conditions

» High humidity and temperatures are necessary for the development of fungal lesions.

Survival and spread

» Infected tissues release conidia, which are then carried by the wind, rain, or insects.

Management

» Gather diseased fruit and throw it away.

» Remove diseased branches and throw them away.

» Eliminate or subdue thrips (wounds caused predispose the plants for infection).

» The best times to apply Copper fungicides are right after the fruit has set, three to four weeks after the main bloom period ends, and when the flower buds first begin to open.

References

Byrne, F. J., Urena, A. A., Robinson, L. J., Krieger, R. I., Doccola, J., & Morse, J. G. (2012). Evaluation of neonicotinoid, organophosphate and avermectin trunk injections for the management of avocado thrips in California avocado groves. Pest management science, 68(5), 811-817.

Dreistadt, S. H. (2007). Integrated pest management for avocados (Vol. 3503). UCANR Publications.

Efendi, D., & Litz, R. E. (2003). Cryopreservation of avocado. In Proceedings V Wor Avocado Congress (Actas V Congreso Mundial del Aguacate) (pp. 111-114

Mutembei, M. M. (2009). Agronomic practices and postharvest management anthracnose in avocado (Doctoral dissertation, University of Nairobi).

Sonavane, P. S., & Venkataravanappa, V. (2022). Avocado (Persea Americana Mill Diseases and Their Management. Diseases of Horticultural Crops: Diagnos and Management: Volume 1: Fruit Crops, 71.

Zentmyer, G. A. (1984). Avocado diseases. International Journal of Pest Managemer 30(4), 388-400.

5

Banana

INSECT PESTS

Rhizome weevil, *Cosmopolites sordidus*

Damage symptoms

- » There are two stages of feeding on the plant: adults and grubs.
- » Weevil grubs eat the rhizome, but adult weevils prefer the phoney stem.
- » Each egg is placed single on the shoulder blade (above ground or on rhizomes underground).
- » The grubs then begin tunnelling into the rhizome shortly after hatching.
- » These crevices are used for the pupation process.
- » Because of the monsoon, the plants become sickly, decompose, and die.

Pest identification

- » **Eggs** - Single, white eggs are laid on the rhizome's upper surface.
- » **Grub** - Apodous, whitish-yellow, and topped with a crimson cap
- » **Pupa** - white in colour, and they develop within the corm through tunnelling.
- » **Adult** - weevils are black, while newly hatched ones are reddish brown.

Management

- » Plant only rhizomes and suckers that are healthy and free of disease.
- » Avoid bringing in a new crop into an infested field by not using it for your usual crop.
- » The rhizome is trimmed to get rid of any diseased or infected parts.

» The varieties Poovan, Kadali, Kunnan, and Poomkalli are more resistan and should be grown.

» The suckers should be washed and dipped in a Chlorpyrifos 20 EC @ 2.. ml/L solution before being planted.

» Before planting, sprinkle 80g of Neem cake, 50g of Carbaryl dust, or 10, of Phorate into each hole.

» Adhere to hygienic cultural norms and keep the orchard clean.

» In the event of an infestation after planting, saturate the soil around th plant's base with Chlorpyriphos 2.5 ml/L and then apply Malathion ml/L a week later.

» Beauveria bassiana, an entomopathogenic fungus, is a potent

» The entomopathogenic nematode *Steinernema carpocapsae* was put to pseud stem traps, and it resulted in a mortality rate of up to 60% for adult weevils

» Set up a 5/ha Cosmolure trap

Pseudo stem borer, *Odoiporus longicollis*

Damage symptoms

» Neither the adults nor the grubs of this curculionid weevil can surviv without a plant's nutrition.

» There, the female lays her eggs in a little air-filled chamber made by cutting the leaf sheath.

» After hatching, the grubs tunnel through the pseudo stems, eating the tissue around the air chambers.

» The exterior surface of the pseudo stem is pierced by grubs.

» The initial sign is the gushing of plant sap.

» Out of the hole comes a dark mass.

» It looks like the plant is wilting.

» Fake stems become brittle and waste away.

» Similar to rhizome weevil, severe winds eventually cause these trees to fal down.

Pest identification

» **Eggs** – which are yellowish white and cylindrical in shape and laid at random on the severed ends of the pseudo stem — are laid in this manner.

» **Grub** - Apodous, whitish-cream in colour with a dark brown top.

» **Pupa** – a pale-yellow, fibrous cocoon that develops inside the tunnel's wall.

» **Adult** - Strong, adult weevils that are rusty brown and jet black.

Management

» Reduce the prevalence of pests by using pest-free, healthy suckers.

» This pest can be avoided by practising clean agriculture.

» Maintain a clean field by regularly raking off the dead leaves.

» Remove the suckers every month.

» In the event of a significant infestation, the plants should be uprooted and burned.

» Apply Chlorpyrifos 20 EC at a rate of 2.5 ml/L twice or thrice each week.

» If the infestation level is modest, injecting the drilled hole with Dichlorvos 0.25 % may be sufficient.

» Ten ml per ml of Azadirachtin is injected into the stem, just above the final hole in the fake stem.

» Use a suspension of *Steinernema feltiae* (5 nematodes/ml) to biologically manage stem weevil grubs.

» Use beneficial pathogens as a form of biocontrol, such as *Beauveria bassiana* and *Metarrhizium anisopliae*.

Aphid, *Pentalonia nigronervosa*

Damage symptoms

» Aphids tend to cluster at the leaf's underside, where they do relatively little damage.

» There is a progressive shrinking and curling of the leaves in the case of a severe infestation.

» The fruit bunches shrink, and the fruits themselves may become misshapen.

» The leaves are arranged in a rosette shape

» The edges of the leaves are undulating and rolled upward.

» There has been a slowdown in the plant's development.

» Don't make piles of anything.

» Banana bunchy top viral transmitter.

Pest identification

» **Nymphs** - round to slightly elongated with a ruddy brown colour and six-segmented antennae.

» **Adult** – Aphid adults are about the size of a grain of rice and can range in colour from red to dark brown to practically black. They have dark, pronounced veins and antennae with six segments.

Management

» Produce only in a pristine environment.

» Reduce the prevalence of pests by using pest-free, healthy suckers.

» Before spraying, use rhizome to remove any infected plants.

» At 2-ml/L intervals, spray either dimethoate (30 EC) or oxy-demeton (25 EC).

» Aim the spray at the top and bottom of the pseudo stem, all the way down to the ground.

» Using 1 ml per plant, inject Monocrotophos 36 SL (1ml diluted in 4 ml of water).

» Don't inject Monocrotophos after flowering, please.

» *Scymnus, Chilomenes sexmaculatus, Chrysoperla carnea*, and other coccinellid predators should be encouraged to do their thing.

» Aphids can be easily controlled by releasing predators like ladybird beetles and lace wings onto the field.

» Apply the entomopathogen *Beauveria bassiana*.

Cut worm, *Spodoptera litura*

Damage symptoms

» The young larvae eat by scraping the leaves off the underside of the adult's body.

» Later, when it's dark, they gorge themselves on the plants.

Pest identification

- » **Larva** - drab, greenish brown with contrasting black markings.
- » **Adult** - The submarginal portions of an adult's eyes are spotted with yellow and purple. Head and Shoulders Moth, the brown area of the forewing is marked with white waves. White with a brown splotch on the margins; these are the wings that flap in the back.

Management

- » Gather up the broken pieces of the plant and throw them away.
- » Use summer ploughing to expose pupae to the elements.
- » Set up a 1-hectare light trap.
- » The use of a spray containing Azinphosethyl, Chlorphyriphos, and Monocrotophos.
- » Do not stop applying Bt even if you see a few bugs here and there.
- » Avanthe 1 ml in 100 ml water should be sprayed on the plant's leaves.
- » Parasitoids like the *Telenomus spodopterae* and *T. remus* are released in the wild to eat eggs.
- » Spreading the entomopathogenic fungus *Nomuraea rileyi* in the wild.
- » Use 250 LE/ha to spray SINPV.

Fruit rust thrips, *Chaetanaphothrips signipennis*

Damage symptoms

- » The banana rust thrips feed on the skin in the space between the banana's fingers, leaving behind rusty discoloration.
- » Fingers stained a rusty crimson colour
- » Leaves turning yellow and a rusty development on the fruit.
- » A decrease in fruit's marketability due to exterior browning.
- » Banana slugs lay their eggs in the peels of bananas and feed on the developing flower buds.
- » Where infestations are especially bad, this might cause a significant drop in fruit production.
- » The skin may develop longitudinal fissures, and the fruit may break, in extreme circumstances.

Pest identification

- » **Larva:** tiny, whitish version of the adult, but otherwise are an exact miniatu of the latter.
- » **Pupa:** White and around the same size as the larvae, the pupae are able to crav
- » **Adult:** The mature moth is only 1.5 mm in length and is either a pa yellow or golden brown colour with fine, feathery wings.

Management

- » Burn down any straggler plants or abandoned plantations.
- » Plant from suckers that are in good condition and free of pests.
- » Treatment with hot water before planting.
- » The early application of bunch covers (which extend the length of the bunc
- » In order to prevent any spoilage, it is crucial to check the fruit under th bunch coverings on a regular basis.
- » Chlorpyrifos should be applied to the bunches, pseudo stem, and sucker
- » Fipronil and bifenthrin sprays for the soil.
- » Placing predatory coccinellids like lacewings and ladybird beetles in the wil

Hard scale, *Aspidiotus destructor*

Damage symptoms

- » Fruits, pseudo-stems, and leaves are all susceptible to attack from grub and adults.
- » Leaf yellowing is occurring in isolated areas.
- » Growth retardation in vegetation.
- » The rhizome was eaten by grubs, killing the plant.
- » The rhizomes have dark, winding passageways.
- » Unopened pipe dies and its outer leaves wither.

Pest identification

- » **Nymph** - oval in shape and transparent, yellowish brown in colour, ar coated in a waxy substance.
- » **Adult** - Female adult ovaries are round, translucent, and light brown.

Management

- » The infected plant components must be gathered and disposed of.
- » Use a spray bottle to apply 0.04% monocrotophos 36 WSC
- » *Chilocorus nigritus* and *Scymnus coccivora*, two predators of coccinellids, are released into the wild.

Root mealy bug, *Geococcus citrinus, G. coffeae*

Damage symptoms

- » Roots with parasites will have whitish, cottony masses.
- » In these whitish blobs, you'll find both adult females and eggs.
- » Infected plants lose their vitality and their growth rate, and their leaves turn yellow or chlorotic.
- » Banana sucking insects, both adults and juveniles, feed on the sap of lateral roots before settling down at the point where the roots of the laterals join the main root.
- » Leaves turn yellow and thin, plant health declines, and fruit yield drops.

Pest identification

- » The eggs are 0.32 mm in length and 0.15 mm in width and have a pearly white colour.
- » The nymphs looked whitish and mealy after the second instar.
- » The adult female is 1.2–2.0 mm in length and has an elongate oval body that is white and free of waxy border filaments. She also has robust antennae, legs, and a pair of anal lobes that end in spines.

Management

- » Seedlings should be boiled for 10-30 seconds before being planted.
- » An entomopathogenic fungus called *Beauveria bassiana* PPRC-56 was responsible for 54 % of the deaths.
- » *Millettia ferruginea* (10% water suspension) applied to the root zone of a sucker.
- » High root mealy bug mortality was observed following the application of Chlorpyrifos 48% EC and Diazinon 60% EC at 1.7 L of solution (after diluting the insecticide in 1:5 L of water) per field-grown plant and poured on the root collar area.

» The root mealy bugs *Geococcus coffeae* and *G. citrinus* that infest the Nendran banana variety were found to be naturally infected with the entomopathogeni fungus *Hirsutella* sp. (Fungi Imperfectii).

Spiraling whitefly, *Aleurodicus dispersus*

Damage symptoms

» Whiteflies prefer to dine on the undersides of plants.

» Whiteflies are harmful because they pierce leaves and feed on the sap, which causes the damage.

» When the whitefly population is high, this results in an early death.

» Mold grows on the sugary secretions made by whiteflies and their nymphs giving heavily afflicted plants a black sooty look.

» When this sooty mould attacks a plant, it weakens or even kills it by preventing it from producing enough sugars through photosynthesis.

Traceability of Pests

» The eggs are a smooth oval shape and range in colour from yellow to tan A trail of white wax is used to hold them in place in a spiral formation.

» Waxed tufts appear on nymphs and larvae as they develop.

» The adult has two distinct black dots on its forewings. Both sexes measure 2.2 mm in length, while males are somewhat larger at 2.3 mm.

Spread

» The crawling nymphs and the flying adults both contribute to the spread while the latter is more effective at long distances. Long-distance transmission occurs on plants that are transported internationally for horticultural purposes

Management

» Whiteflies in their larval stages are killed by the eggs of the wasp parasitoid *Encarsia haitiensis*. Adult lacewings and ladybird beetle larvae feed on both adults and young.

» Spiral whitefly nymphs are the only host for the biological control agent *Encarsia dispersa*, which has been used effectively to reduce pest populations in Torres Strait and mainland Queensland.

» The predatory insect *Cryptolaemus montrouzieri* preys on spiralling whiteflies in general.

» Use white oil (3 tablespoons (1/3 cup) cooking oil in 4 litres water, 1/2 teaspoon detergent soap, shake well, and use) and horticulture oil (produced from petroleum). soap solution (five tablespoons of pure soap and two tablespoons of dishwashing liquid per four litres of water), etc.

Slugs, *Ariolimax* sp. and snails, *Achatina* sp.

Damage symptoms

» Rasp big holes in plant matter including leaves, stems, fruit, and bulbs, and eat the tender, succulent flesh in between.

» They can do significant damage to young plants and even destroy seedlings.

» Snails and slugs, on occasion, will make the ascent to higher feeding grounds, such as trees and shrubs.

» In wet years or locations with high rainfall, populations of both increase, and harm is greatest.

Management

» Eliminating slugs and snails by hand.

» Cover your garden at night with pots, boards, cabbage leaves, or citrus rinds. Crawling under them for shelter from the sun and heat, snails and slugs will discard them the following morning.

» Place a saucer of old beer or a yeast and honey concoction on the table. Bury it so that the highest point of the saucer is even with the ground. The mollusks and gastropods will drown if they enter the combination.

DISEASES

Panama wilt, *Fusarium oxysporum* f. sp. *cubense*

Symptoms

» The elder leaves will begin to turn yellow at the edges, and eventually the midrib will follow suit.

» Browning, drying, and eventual falling off of the leaves.

» Eventually, only the youngest leaves will remain green and upright, while the older ones will curl around the stem to produce a "skirt."

» The leaves will fall off the trees eventually.

» The stems of certain species can divide.

» Brown, red, and yellow rings appear inside the stem, starting in the middle and then extending outward in the case of a serious infection.

» Even suckers can get sick sometimes.

» Everything, above and below earth, eventually dies and rots.

Management

» Six months of flooding in famine-stricken communities.

» Use rice or sugarcane in a crop rotation.

» Two, four, and six months after planting, apply the capsule containing Pseudomonas fluorescens/Carbendazim 50 WP at a dose of 60 mg/capsule.

» Compare and contrast the banana. There have been reports of wilt-resistant varieties of giant cavendish, lacatan, rajavazhai, peyladen, moongil, poovan and vamanakeli.

Sigatoka leaf spot, *Mycosphaerella fijiensis*

Symptoms

» Reddish-brown streaks, around 1-5 mm long and 0.25 mm wide, that run parallel to the veins are the first visible evidence of the disease.

» On the underside of the third or fourth youngest leaf, particularly around the edge of the leaf blade, they are most obvious.

» The expansion and visibility of the streaks on the upper surface is accompanied by a darkening of their centres, a transition to a grey color, a little depression, black perimeters, and a dazzling yellow halo.

» The leaves eventually wilt and fall off as the streaks coalesce into bands of death a few mm broad on either side of the midrib.

Management

» Avoid thick clay soils where surface water persists for an extended period of time after rainfall and instead grow in well-drained regions.

» A lower disease rate can be achieved through preventative measures such proper spacing, appropriate drainage, weed control, and the elimination of suckers.

» Infected leaves must be plucked and burned.

» When disease pressure is low, spray protectant fungicides such as Mancozeb (in oil or oil/water emulsion) and Chlorothalonil (in water).

- » Systemic fungicides, such as the Triazoles (including Propiconazole, Fenbuconazole, and Tebuconazole), and the Strobilurins, should be sprayed (e.g. Azoxystrobin).
- » Spraying a mixture of 1% Bordeaux wine and 2% linseed oil.
- » Field testing showed good results from spraying with 0.1% Topsin-M or Prochloraz or 0.1% Carbendazim or 0.2% Chlorothalonil 0.15% or Kitazin three to four times every two weeks.

Anthracnose, *Gloeosporium musarum*

Symptoms

- » The fungus typically infects the mature end of immature banana fruits.
- » In the first stage, infected fruits grow tiny black spots in a circular pattern.
- » These spots eventually grow in size and darken to a brown.
- » In this stage, the fruit's skin goes black, dries off, and becomes covered in tiny, pink fruiting bodies (acervuli).
- » The whole finger is impacted at the end.
- » The disease eventually spreads throughout the whole bunch, causing the fruit to shrivel and turn a pinkish colour from the spores.
- » When black lesions form on the pedicel, the pedicel withers and the fingers fall off the hands.
- » Unfortunately, the bunch's central stalk might sometimes get ill.
- » Fruits that have been infected will turn a dark, rotting colour.

Pathogen identification

- » In terms of shape, acervuli can be either spherical or elongated and erumpent.
- » Conidiophores are elongated and narrowing towards the apex; they are hyaline and septate and branching and sub-hyaline at the base; and they have a single phialidic aperture at the very tip.
- » Hyaline, aseptate, oval, ellipsoid, or straight cylindric conidia with obtuse apices or flattened bases and obtuse apices are the hallmarks of these guttate spores.

Spread

- » Conidia in the air and a wide variety of insects that feed on banana blooms are responsible for the rapid spread of this disease.

Management

» When all the hands have opened, it's important to remove the distal bud to avoid spreading infection in the field.

» The diseased plant material should be burned.

» Keep the playing field clean.

» The disease can be stopped by spraying young bunches with a Bordeaux mixture diluted to 1 %.

» Highly effective is spraying the crop three to four times at two-week intervals with either 0.2% Prochloraz, 0.1% Carbendazim, 0.2% Chlorothalanil, or 0.15% Kitazin before harvest (Rawal and Ullasa, 1989).

» To avoid fruit rot after harvest, fruits can be dipped in a solution containing 440 ppm Mycostatin, 100 ppm Aureofungin, 400 ppm Carbendazim, or 1000 ppm Benomyl.

» Banana bunches should be picked at the ideal maturity level and kept at a cool 10 °C until use.

Cigar end rot, *Verticillium theobromae*

Symptoms

» Both ripe and unripe fruits are susceptible to the disease.

» Some bunches only have a few infected fruits, while others have the disease across all of the fingers.

» Blackening, shrinking, and folding of the skin tissues are the results of the infection that begins in the perianth and progresses down the finger.

» Fungal conidiophores and spores, known collectively as "conidia," begin to blanket the afflicted corrugated area.

» Cigar end disease gets its name because of how similar the symptoms are to the ash left on the end of a cigar.

» Most rotten fruit is shorter than 2 cm.

» However, in cases of serious diseases, decay can be seen from the very tip of the fruit.

» Dry rot occurs when the interior pulp dries up and becomes fibrous.

Management

» In order to lessen the likelihood of disease transmission inside a banana plantation, it is important to eradicate any other *Musa* spp., *Bamboo* spp., *Heliconia bethel*, or *H. resiliencies* that may be present.

» Disease can be kept at bay by removing the fruit's perianth and pistil by hand as soon as possible after it forms 8-11 days after a bunch has emerged is the optimal time to remove the pistils.

» When it's rainy outside, it's especially important to remove the bracts that are still attached to a young bunch and expose it to the light and air.

» Plants should not be crammed too closely together in the plantations; this will ensure adequate air circulation.

» Additionally, cleaner public spaces help cut down on the spread of disease.

» Polythene sheets placed over the stems prior to hand emergence can successfully control the disease.

» Wetting agent at a concentration of 0.5–1.0 ml/L of spray fluid and COC at 0.25% can be sprayed on the bunches.

Bacterial wilt, *Ralstonia solanacearum*

Symptoms

» Premature ripening of fruits, leaves that wilt and fall off quickly, browning of the vascular threads, and withering, blackening of the suckers are all symptoms.

» Wilt is especially dangerous for young plants because it can spread quickly.

» It may just take a week or less from the first signs to the complete death of the plant.

» The fruit spoils, the fruit stem becomes brown, and the suckers wilt or turn black.

» The lower leaves turn yellow first, and then the rest of the plant follows suit.

» It is not uncommon for the junction of the petiole and lamina to collapse on the second or third leaf, or even on both leaves.

» Sometimes plants die out before the appearance of bunches.

» Its green stem has some yellow appendages.

» The vascular filaments in fruit stalks are similarly stained.

» The fruit may be smaller than usual, and the pulp may exhibit signs of firm dark or grey rot.

Survival and spread

» The most common inoculation method (Primary source of inoculum): The bacterium can travel through the soil and water and can also infect the plant's suckers and rhizomes. The banana and Heliconia species provide suitable hosts for its survival.

» Secondary reservoir of infectious agents (Secondary source of inoculum): In addition to being carried by the water used for irrigation, bacterial cells can also be distributed by the suckers that are utilised in the planting process.

Management

» Make sure the plant material you're using is disease-free.

» Plants like Robusta and Grand Nadine that are resistant to disease should be used.

» Any infected vegetation should be destroyed by fire.

» Drainage must be addressed, and water should not be transferred from diseased to healthy plants.

» The use of drip irrigation has been shown to lessen microbial growth.

» The region needs to be left fallow for 6-12 months, during which time it should be kept weed-free, in order to control the disease.

» Kudzu *(Pereira phaseoloides)* and sorghum are two good cover crops for lowering the soil inoculum.

» Put *Pseudomonas fluorescens* or another bio- agent to use.

Bunchy top, *Banana Bunchy Top Virus* (BBTV)

Symptoms

» On the underside of the leaf, along the midrib and petiole, symptoms manifest as uneven, nodular, green streaks along secondary veins.

» Initial infection at the stool or mother plant stage results in tiny, closely spaced leaves (suckers).

» They don't bend over like they would on a healthy banana plant, but instead remain standing straight.

» Lighter (chlorotic) along the margins, fading to brown in the middle of the leaf on its downward spiral.

» Each subsequent generation of leaves is narrower and smaller than the one before it, giving the plant its characteristically dense and clumpy appearance.

» The next leaves are gradually smaller, more chlorotic at the edges, and more prone to curling.

» Petioles are short and leaves are fragile.

» As the plant's crown develops into a rosette or clump of stunted leaves, the leaves finally die off.

» The emergence of clusters is hindered by the pseudo stem.

» The value of bundles declines as they are sized down.

» Fruitful clusters are a rarity on suckers that have been grafted onto diseased stools.

Spread

» Spread via corms, cuttings, and suckers found in plant nurseries.

» Banana aphids, specifically *Pentalonia nigronervosa,* are the primary transmitters. This virus spreads continuously from person to person.

Management

» These suckers need to come from healthy gardens.

» The infected plants and their rhizomes need to be uprooted and discarded at regular intervals.

» Before planting, sprinkle 40 g of Furadan 3G granules into each hole.

» Apply a fungicide spray containing either 0.1% phosphomidon, 0.2% oxydemeton methyl, or 0.1% monocrotophos on the crop three times, with a three-week interval between each application.

» From the third month till flowering, inject 1 ml of Monocrotophos (1 ml diluted in 4 ml of water) per plant every 45 days.

» From a clone that has been infected, virus-free meristems and plants can develop.

Bract mosaic, *Bract Mosaic Virus* **(BMV)**

Crop losses

>   Banana bract mosaic virus weakens banana plants and causes deformed bunches and underdeveloped fingers, which can diminish bunch weight by as much as 40 %.

Symptoms

>   Infected plants will show symptoms like black streaking on leaves, yellowing or pinking of the pseudo stem, and purple spindle-shaped mosaics on the bracts.

>   A potyvirus is responsible for causing the disease.

>   Even in extreme cases, the midrib's underside retains a pinkish tinge.

>   When suckers emerge and the leaf sheath separates from the central axis the typical reddish brown stripes appear.

>   "Traveler's palm"-like leaf clustering at the crown.

>   Its hallmark characteristics include a prolonged peduncle and partially full palms.

>   The symptoms of this virus disease are readily apparent in banana cvs. Pairs of Monthan and Plantain.

>   Defects in growth, diminished suckering, and abnormally shaped fruit are all possible outcomes of infection.

>   Rejection of fruit and the accompanying economic losses might result from more severe cases.

Pathogen identification

>   Virion filaments are bendable rods.

Survival and spread

>   The aphids *Aphis gossypii, Aphis craccivora, Pentolonia nigronervosa,* and *Rhopalosiphum maidis* are the vectors for this virus.

>   The suckers were the principal vector of disease transmission in the field.

Management

>   Any new plantings should only use disease-free plant material.

» It is important to get rid of the sick plants as soon as possible to stop the disease from spreading.

nfectious chlorosis, *Cucumber Mosaic Virus* (CMV)

ymptoms

» Bands of mosaic-like or discontinuous linear streaking from margin to midrib are diagnostic of this virus.

» Newly emerging leaves stand erect and rigidly twisted and bunched at the top are also noticed.

» Early signs of heart rot, caused by the decay of the heart leaf and the centre region of the pseudo stem due to the interaction of some bacteria, include the appearance of dead or dying suckers.

pread

» The disease typically spreads from plant to plant via the daughter suckers of affected plants.

» It is the *Aphis gossypii* mite, not the *Pentalonia nigronervosa* fly, that is responsible for secondary disease transmission.

lanagement

» Banana plantations should have no weeds growing in them since the virus can overwinter there.

» Infected clumps should not have their suckers used for new plantings.

» Avoid planting cucurbits like pumpkins, cucumbers, and others in between your banana rows.

» Suckers were successfully virus-free after being dried and heated to 40 °C for a day.

» The vector population can be kept in check and disease transmission slowed by spraying 0.03% Methyl demeton every three to four weeks.

» Compare and contrast the banana. The disease cannot infect Ela Vazhi, Athiya Kol, or Karu Bale.

anana streak, *Banana Streak Virus*

ymptoms

» The most typical signs are chlorotic or yellow streaks that run from the leaf midrib to the leaf margin and can be narrow, discontinuous, or continuou Spindle- or eye-shaped patterns can be seen in some circumstances.

» Banana streak has also been linked to spots of yellow.

» You may experience few or many symptoms.

» A deformed lamina is possible.

» Later, the orange streaks become darker, sometimes even brown or black

» The midrib and the petiole have also shown signs of necrosis.

» Lower temperatures and fewer hours of daylight increase the likelihoc of necrosis.

Spread

» Infected seeds and soil are a major vector for the virus's rapid spread.

» Mealybugs such as the citrus mealy bug (*Planococcus citri*), the pineapple mea bug (*Dysmicoccus brevipes*), and the pink sugarcane mealy bug spread the vir from banana to banana in a semi-persistent manner (*Saccharicoccus sacchari*).

Management

» It is necessary to eradicate infected plants.

» Substituted with plants created from virus-free meristems of plants alreac screened for the virus.

» If there is a high prevalence of viruses and the disease looks to be movir from plant to plant, the vectors of the disease, which are mealy bugs, shou be eliminated.

References

Blomme, G., Dita, M., Jacobsen, K. S., Pérez Vicente, L., Molina, A., Ocimati, W ... & Prior, P. (2017). Bacterial diseases of bananas and enset: current state knowledge and integrated approaches toward sustainable management. *Frontie in plant science*, 8, 1290.

Jones, D. R. (2019). Fungal diseases of banana fruit. In *Handbook of diseases of banan abacá and enset* . Wallingford UK: CAB International. pp. 255-295.

Kannan, M., Padmanaban, B., Anbalagan, S., & Krishnan, M. (2021). A revie on monitoring and Integrated management of Banana Pseudostem Weev Odoiporus longicollis Oliver (Coleoptera: Curculionidae) in India. Internation Journal of Tropical Insect Science, 1-9.

Kumar, P. L., Selvarajan, R., Iskra-Caruana, M. L., Chabannes, M, and Hanna, R. (2015). Biology, etiology, and control of virus diseases of banana and plantain. *Advances in virus research*, *91*, 229-269.

Padmanaban, B., & Mani, M. (2022). Pests and Their Management in Banana. *Trends in Horticultural Entomology*, 577-603.

Raut, S. P. and Ranade S. (2004). Diseases of banana and their management. In *Diseases of Fruits and Vegetables: Volume II: Diagnosis and Management*. Dordrecht: Springer Netherlands, pp 37-52.

Viljoen, A., Mahuku, G., Massawe, C., Ssali, R. T., Kimunye, J., Mostert, D., ... & Coyne, D. (2017). Banana diseases and pests: Field guide for diagnostics and data collection. International Institute of Tropical Agriculture (IITA).

6

Ber

INSECT PESTS

Fruit fly, *Carpomyia vesuviana*

Damage symptoms

» When fruit begins to set, the infestation begins.

» Adults prey on ber fruits and deposit eggs within them.

» Adults attack ber fruits and lay eggs inside fruits

» Rotting of the fruit can occur if larval excrement builds up in the galleries

» Fruits that have been infested by a pest will not develop normally and will look abnormal.

» Many of these fruits are lost.

Pest identification

» **Egg:** A typical chicken egg is white, oval, and about 1/25th of an inch long and wide.

» **Larva:** In its larval stage, the white larva looks like a skinny cone and has no legs. The mouth can be found at the snout. An instar 3 is roughly half an inch in length.

» **Pupa:** Puparium, the pupae's food, is a seedlike yellowish-brown fruit

» **Adult:** Two black stripes run horizontally across an adult's belly, while a third, longer stripe runs from the third segment's base to the abdomen's tip

Favorable conditions

» Regular weekly precipitation of 20-40 mm during July and August stimulated adult activities.

» Pests thrived in conditions with a mean weekly minimum of 10-15°C, a maximum of 25-40°C, and a relative humidity of 25-90%.

Management

» To find pupae, one must dig around the roots of trees.

» All bruised fruit must be gathered and thrown away.

» Make use of traps with 1 ml of methyl eugenol and 2 ml of malathion per litre of water. Use 100 cc of the solution per trap and space them out so that there are 10 traps per hectare.

» It is possible to reduce insect incidence by picking fruits when they are still green and firm to prevent the pests from laying eggs on them.

» Malathion sprays with a concentration of 0.6% were used to mitigate the harm.

» Parasitoids come in many forms, such as *Fopius arisanus* and *Diachasmimorpha kraussi*.

» In term of berries, the varieties Sanaur-1, Safeda Selected, Ilaichi, Mirchia, ZG-3, and Umran fare better against fruit flies than others.

Fruit borer, *Meridarchis scyrodes*

Crop losses

» In extreme cases of infestation, the bug can cause a loss of production of up to 70%. The level of destruction could be somewhere between 20% and 80%.

Damage symptoms

» The moths lay their eggs on fruits when they are in the pea stage, and the freshly hatched caterpillars bore into the fruits, feeding on the pulp close to the seed and accumulating faeces as they go.

» Earlier stages of larvae (instars 1 and 2) just eat the fruit's outer layer, but later stages (instars 3–5) eat their way into the fruit and cause harm to the pulp surrounding the seed.

» In the early phases of fruit growth, full-grown larvae are discovered feeding on the tender immature seed.

Pest identification

- » The first instar larvae are a pale yellow, whereas the mature larvae are a vibrant red.
- » The wings of an adult fly, which is tiny and dark brown, are fringed.

Management

- » Getting rid of maggots and pupae in the soil requires collecting and destroying fallen fruit and digging orchard soil under the tree canopy.
- » Repeatedly rake the ground.
- » The percentage of borer damage was lowest in areas sprayed with Neem seed kernel extract at 5%.
- » Initially, spray with 0.03% Monocrotophos at the pea stage, then reapply 15 days later with 0.05% Fenthion, and finally, repeat the process with 0.01% Carbaryl 15 days after the second application of Fenthion.
- » Before the fruits have reached the marble stage, spray 40 kg of Chlorpyriphos 1.5 D per hectare around the plants.
- » At fruit set, spray Malathion or Dimethoate with a rate of 1.0 litres per litre (L) twice at a 15-day interval between applications.
- » There are two known borer parasites, *Microbracon* sp. and *Opius carpomyiae.*
- » Resistance is seen in the Cvs. Gurgaon, Banarsi Pewandi, Ajmeri, Gola, and Jhajjar Selection varieties. Fruit fly resistance was higher than moderate fruit borer resistance in Cv. Ilaichi.
- » Culture (hoeing + collecting of fallen fruits from June to December) and pest control (insecticide (Dipterex @ 250 g/ha)) make up the integrated management approach. was deemed to be the most useful.

Bark eating caterpillar, *Indarbela quadrinotata*

Damage symptoms

- » Caterpillars graze at night on the tree's bark.
- » They make tunnels inside the trunk or major stems, eating the wood from the inside out.
- » Huge silken webs blanket the damaged area.
- » There is a risk of plant death if the infestation is severe.

Pest identification

» **Larva:** Larvae range in colour from dark brown to black; they are 4.5 to 5.0 cm in length, have a glossy appearance, and have only a few short hairs.

» **Adult:** The adult moth is a creamy white colour with brown patterns on the forewing.

Management

» Take off the peeling galleries and cover the bark in paint.

» In order to properly control the bark-eating caterpillar, a solution containing 1 litre of kerosene, 100 g of soap, and 9 litres of water should be applied to the holes.

» Using a cotton swab dipped in gasoline or kerosene, wipe down the damaged area.

» With a syringe, inject 5 ml of dichlorvos into the borehole, and then fill it in with mud.

» Carbofuran 3G granules, at a rate of 5 g per bore hole, followed by a mud plug should be used.

» Swab the trunk with 20 g/liter of Carbaryl 50 WP or pad with 10 ml of Monocrotophos per tree.

Lac insects, *Kerria lacca, K. indica*

Crop losses

» A loss of 52.5% to 58.5% of fruit was produced due to an infestation of 5000 nymphs/100 cm twigs.

Damage symptoms

» Damages the tree's vitality and severely reduces its ability to bear fruit.

» Horribly sucking insect that will kill a tree by sucking all of the life force out of its branches.

Management

» At the time of the annual trimming, any infected branches should be removed and thrown away.

» A second round of treatment is possible in around three months if necessary.

» Spraying trees with 0.1% Dimethoate or 0.03% Phosphamidon after pruning is recommended.

» Many different kinds of wasps, two kinds of moths, and three kinds of lacewing flies all prey on the lac bug.

DISEASES

Powdery mildew, *Oidium erysiphoides* f. sp. *zizyphi*

Crop losses

» It causes a loss of 50-60% of fruit yield and a consequent drop in price at market.

Symptoms

» White powdery spots on young leaves are the first sign of this disease, which then moves on to the developing fruit, flower buds, and new shoots.

» Discoloration shifts from yellowish to varying shades of brown on spots.

» This issue causes the leaves to fall off too soon.

» Roughness and mild elevation occur in the infected area.

» Malformed and flavourless fruit is a common result of infection.

» Continued cold weather causes fruits to become cankered and cracked.

» Yield and market quality decline.

» The dropping of immature fruit.

Favorable conditions

» Intolerable humidity.

» We can expect cloudy skies, temperatures between 10 and 300 °C, with an infrequent chance of precipitation.

» RH more than 90%.

Survival and spread

» The host plant's bud wood is where it survives.

» Conidia in the air cause a secondary infection to spread.

Management

» Pruning is done between the second and first weeks of May.

» Apply 150–200 g of sulphur per tree, then repeat the process three more times at intervals of 15–20 days.

» In September (during blooming), in the middle of October, in the middle of November, and in the middle of December, spray 0.25% Wettable Sulfur (250 g in 100 litres of water) or 0.05% Karathane 40EC (50 ml in 100 litres of water) or 0.05% Bayleton 25 WP (50g in 100 litres of water). In January, if necessary, a second spray might be administered.

» Tridemifon, Benomyl, or Carbendazim 0.1% Spray.

» The following cultivars are resistant to the virus: Cv. Safeda Rohtak, Sua, Noki, Chonchal, Saunar 1 & 5, Katha Phal, Illachi Jhajjar, Kakrola Gola, Kala Gora, Pathani, Jogia, Mundia, and Mirchia.

Leaf spots, *Alternaria alternata, Cercospora ziziphi*

Symptoms

» Small irregular dark brown spots appear on the upper leaf surface as a symptom of the disease.

» Spots of a dark brown to black colour appear on the underside.

» Massive patches emerge as the spots join together.

» Yellowing and eventual drop characterise the damaged leaves.

» Brownish, indented, elongated dots are commonly found on fruits.

Favorable conditions

» A humid climate.

» Constant precipitation.

Survival and spread

» The original infection may have originated on the plant detritus.

» Wind-borne conidia have a secondary role in the disease's dispersal.

Management

» Mancozeb or Carbendazim at 0.2% is all that is needed to properly manage the condition.

» Spraying with Dithane M-45 (0.25%) or Foltap (0.1%) as soon as the disease emerges is an effective way to control either of these diseases.

» The disease's severity will determine how often you should spray after the initial application, but often every 15 days to 20 days.

Black leaf spot, *Isariopsis indica* **var.** *zizyphi*

Symptoms

>   » On the undersides of leaves, black specks resembling soot tufts appear.

>   » As the disease spreads, it first appears as a brown spot on the underside o the leaves, and then spreads to the upper surface.

>   » As a result, the leaves defoliate and lose their turgidity and shine.

Pathogen identification

>   » The long, dark brown conidiophore produced by this pathogen is multi septate and covered in conspicuous scars

>   » Conidia ranged in size from 17 to 42 × 8.5 to 10.2 μm and were a tan t light brown colour. They were multi-celled (3-4 cells), broader in the middl and narrowed at the ends.

>   » A list of conidia that germinated from the tip cell was also compiled.

Favorable conditions

>   » October and November provide cloudy skies and mild temperatures.

>   » Dry air is ideal for the spread of the disease.

Survival and spread

>   » Fungal infections typically originate in the soil or plant matter.

>   » Conidia in the air can cause secondary infections.

Management

>   » Beginning in the first week of September, when the disease first appears spraying with either Propiconazole (Tilt) at 0.1%, Difenconazole (Score at 0.1%, or Carbendazim at 0.2% every 15 days is recommended.

>   » The Bahadugarhia, Govindgarh Special, Gola Gurgaon, Popular Gola, an Seo cvs are all resistant to a wide range of conditions.

Anthracnose, *Colletotrichum gloeosporioides*

Symptoms

>   » Spots of reddish brown colour with a yellowish edge and a size of 2 to mm in diameter were noted as a symptom of the disease on affected leave

> » These dots are not limited to the veins of the leaf and can appear anywhere on the upper surface.

> » These spots likely began as individual lesions on diseased leaves but eventually joined together to form larger patches.

> » Small, brown or black patches with a depressed sporulating zone in the centre appeared on the surface of the fruit as an indicator of a disease.

> » On average, there were three to four spots seen on each fruit, and their size increased rapidly as the fruit's peel changed from green to yellow.

Pathogen identification

> » Large numbers of single-celled spores are released by the pathogen onto the surface of the host.

> » The nearly hyaline conidia are bullet-shaped and are formed on short, upright conidiophores by an acervullus that lacks specialised hyphae called setae.

Favorable conditions

> » During the summer, ber anthracnose was noticed.

Survival and spread

> » Since it can feed on dead organic matter in the soil, the pathogen can remain there for a very long time.

> » As a result, this becomes the principal vector of disease.

> » Rain droplets spread the spores through the air and serve as a further source of infection.

Management

> » The most reliable approach of control is to use transplants from individuals who have never been exposed to the disease.

> » With drip irrigation, the spread of the disease is reduced.

> » Low fertility circumstances make plants more resistant to disease.

> » Disease could be mitigated by either decreasing nitrogen rates or switching to nitrate-based nitrogen sources rather than ammonium-based ones.

Rust, *Phakopsora zizyphi-vulgaris*

Symptoms

» The uredopustules, which are small and irregular in shape and a reddish brown colour, first formed on the undersides of the leaves and then spread to cover the entire leaf surface.

» The upper surface of the leaf stays green while the underside becomes yellow

» Finally, the diseased leaves fell from the tree.

Pathogen identification

» The uredospores, which are produced by the pathogen in high numbers inside the lesion, range in shape from circular to oval and are a light brown colour

» Ber leaf rust was only seen during the months of February and May, the drier months of the year.

» The spores formed at the end of the tiny spore-bearing structures inside the pustules that broke the skin's surface.

Management

» Apply a spray of either 0.2% Mancozeb, 0.2% Zineb, or 0.2% Wettable Sulfur.

» It has been reported that the following cultivars are resistant to cv. banarasi sanar-1, sanar-2, sanar-3, kishmish, narma, safeda, and sanar-2 from the safeda selection.

Soft rot, *Phomopsis natsume*

Symptoms

» The disease manifests as an uneven patch on the fruit, with a light reddish vinaceous tint.

» It causes the fruit to swell and becoming pulpy, brown to black in colour with a pliable, flimsy skin.

Management

» Spraying with 0.05% Carbendazim is effective for disease control.

References

alikai, R. A. (2008). Insect pest status of ber (Ziziphus mauritiana Lamarck) in India and their management strategies. In *I International Jujube Symposium 840* (pp. 461-474).

Haldhar, S. M., Deshwal, H. L., Jat, G. C., Berwal, M. K., & Singh, D. (2016). Pest scenario of ber (Ziziphus mauritiana Lam.) in arid regions of Rajasthan: a review. Journal of Agriculture and Ecology, 1, 10-21.

Haldhar, S. M., Mani, M., & Saroj, P. L. (2022). Pests and Their Management in Ber (Ziziphus mauritiana). *Trends in Horticultural Entomology*, 783-801.

Jamadar, M. M., Venkatesh, H., Balikai, R. A., & Patil, D. R. (2008, September). Forecasting of powdery mildew disease incidence on Ber (ziziphus mauritiana lam.) based on weather. In *I International Jujube Symposium 840* (pp. 447-454).

Karuppaiah, V., Haldhar, S. M., & Sharma, S. K. (2015). Insect pests of Ber (Ziziphus mauritiana Lamarck) and their Management.'. *Insect pests management of fruit crops*, 271-294.

Nizamani, I. A., Rustamani, M. A., Nizamani, S. M., Nizamani, S. A., & Khaskheli, M. I. (2015). Population density of foliage insect pest on jujube, Ziziphus mauritiana Lam. Ecosystem. *Journal of Basic & Applied Sciences*, *11*, 304.

Pataraddi, A. R., Jamadar, M. M., & Balikai, R. A. (2008). Status of diseases on ber (Ziziphus mauritiana Lamarck) in India and their management options. In *I International Jujube Symposium 840* (pp. 383-390).

Yadav, S. M., Sharma, V. K., & Sharma, P. K. (2020). Ziziphus mauritiana L.: An overview. *Tropical Journal of Pharmaceutical and Life Sciences*, *7*(1), 01-18.

7 Citrus

INSECT PESTS

Leaf miner, *Phyllocnistis citrella*

Damage symptoms

- » Silvery serpentine mines are made by the larvae as they munch on the epidermis of young leaves and blossoms.
- » The disease causes the leaves to wilt and fold in on themselves.
- » Tender shoots may also have their epidermis mined by the larvae.
- » Defoliation may occur if an infestation is severe enough.
- » Contribute to the spread of citrus canker.
- » Adults and juveniles alike feed on leaf sap.
- » Decline and fall of flowers.
- » Diseased leaves have a cupped, crinkled appearance.
- » As a result, plant development is stunted.

Pest identification

- » **Eggs:** a little, rounded object located just below the midline.
- » **Larvae:** Tiny, yellow or red, apodous creatures that live on the folded leaf margins.
- » **Adult:** A little moth with a black patch on its leading edge as an adult.

Management

- » Set up sticky yellow traps.
- » Apply a 4% solution of Neem seed kernel extract during the time of new flush emergence if infection levels are low to moderate.
- » Spraying an early-stage infestation with a mixture of one part fish oil resin soap to one part Nicotine sulphate to fifty parts water.
- » Sandwich sprays of Neem seed kernel extract 4%, Cypermethrin 25 EC @ 0.5 ml/l, and Neem seed kernel extract 4% should be applied every two weeks during periods of high infestation.
- » *Mallada boninensis* (30 larvae/tree) and *Tamarixia radiata* (40 adults/tree) are released together.
- » *Tetrastichus phyllocnistoides* and *Ageniaspis* sp. release.

Black fly, *Aleurocanthus woglumi*

Damage symptoms

- » Adults typically deposit eggs in a spiral pattern on young leaves.
- » The black nymphs feed on plant sap and drain its vitality.
- » Withered leaves that have curled inward.
- » The leaves drop prematurely.
- » Sooty mould fungus growth is aided by honey dew excretion.
- » The leaves' photosynthetic activity is disrupted and they turn dark.
- » The affected trees bloom poorly and bear tasteless fruit.

Pest identification

- » **Nymphs:** scaly in appearance and have a flattened oval form.
- » **Adult:** tiny insects that are black and lustrous with a light coating of grey all over their bodies. The tip of its abdomen is being covered by wings.

Management

- » Keep water away and keep your plants spaced out.
- » Cut off the infected branches and dispose of them together with any nymphs, pupae, or adults you find.
- » Reduce your use of nitrogen fertiliser and water.

» Neem seed kernel extract 4%, which has no effect on beneficial insects, ca be sprayed.

» Apply 2 ml/L of Chlorpyriphos or Dimethoate (30 EC) at each subsequen flush, waiting 10-12 days between applications. You should spray the solutio all the way up into the plant's canopy.

» Parasitoids, such as *Amitus hesperidium* and *Encarsia clypealis,* can achiev up to 90 % suppression.

» Predators Also, 10-15 eggs/1st instar larvae/plant of *Chrysoperla cornea* o *Mallada boninensis* can be set free.

» There was a 28-30% decrease in the black fly population after the discharg of *M. boninensis* (30 larvae/tree) and *Tamarixia radiata* (40 adults/tree).

» Natural enemies such as *Encarsia sp., Eretomocerus serius,* and *Chlysoperl* sp. should be actively supported.

Lemon butterfly, *Papilio* spp.

Damage symptoms

» Larvae are particularly destructive since they feed on a lot of citrus plant leaf at once.

» Caterpillars feast voraciously on fragile, light-green leaves, leaving just th midribs behind.

» The pests are most prevalent in nurseries and on young trees, and they ar active all but throughout the winter.

» Young citrus trees (those less than 2 feet tall) are especially vulnerable t the caterpillars, and citrus nurseries can be wiped out if enough of then are allowed to survive.

» Caterpillars may like the new foliage and leaf flush of mature trees. Th onset of a new flush is when things pick up the most momentum.

» When an infestation is bad enough, the entire tree loses its leaves.

Pest identification

» The eggs are tiny, white, and spherical.

» The caterpillar is primarily yellowish green with a few oblique brownisl stripes. The animal's dorsal side sports a horn-like feature.

» The adult stage of this species is characterised by a huge, stunning butterfly with green wings that are spotted black.

Management

» Larvae, which first appear as what appear to be bird droppings, must be collected and disposed of.

» New flushes can be treated with a spray of 1 ml/L DDVP (Nuvan) or 3 g/L Carbaryl 50 WP.

» Spread out 500 mature *Trichogramma chilonis* per tree.

» Protect native populations of *Distatrix papilionis*, a parasitoid that feeds on the larvae of braconid pests.

» Set free the parasitoids *Trichogramma chilonis* and *Daphnia papilionis*.

» To treat with *Bacillus thurungiensis* var. *kurstaki*, spray 0.5 g of active ingredient per litre (single application for each generation).

» Parasitoids like *Distatrix (=Apanteles) papilionis* and *Telenomus* species may be responsible for as much as 73% of egg and larval parasitism in the braconid genus.

Aphids, *Toxoptera citricidus, T. aurantii*

Damage symptoms

» Eat young plants and blossoms.

» Adults and juveniles alike deplete plants' vitality by sucking their sap from leaves and stems.

» Flowers withering and falling to the ground.

» Leaves with insect damage have a cupped, crinkled appearance, and developing fruits rot and fall off early.

» As a result, plant development is stunted.

» Sooty mould thrives on the sugary solution secreted by the bug.

» As transmitters of the tristeza virus, aphids play a crucial role.

Pest identification

T. citricidus

> » *T. citricida* adults are glossy black, while nymphs are drab grey or reddish brown

> » Female adults (alata) measure 1.1–2.6 mm in length with six-segmented antennae.

> » Adult females (aptera) measure between 1.5 and 2.8 mm in length and have an oval shape.

> » More than 20 hairs can be found on the cauda.

T. aurantii

> » *Toxoptera aurantii* are 2 mm in length, oval in shape, and either shiny black brownish black, or reddish brown in colour. Their antennae are black and white banded but relatively short.

> » Adults with wings are glossy and dark brown to black, whereas those without are glossy and jet black.

> » There are typically less than 20 hairs in the cauda.

Management

> » Set up sticky yellow traps.

> » Metasystox (Methyl demeton) or Rogar (Dimethoate) 2 ml/L spray.

> » Spray Imidacloprid 200 SL at 0.25 ml/L if an outbreak becomes severe.

> » Natural predators can control the remaining aphid population.

> » In the absence of predators, release 50 *Cheilomenes sexmaculata* per plant.

Mealy bugs, *Planococcus citri, P. lilacinus*

Damage symptoms

> » White mealy bug colonies form at branch intersections.

> » When nymphs and adults alike feed on the sap of young branches, the fruits lose their vibrant color and wilt.

> » Curled leaves are the result.

> » The affected areas of the plant wither and dry out, and the tree's overall vitality is diminished.

> » Honey dew is secreted in great quantities, which can lead to the growth of

sooty mould. The fruit and the leaves are completely infected with fungus.

» When pests are at a bad enough level, blooms won't set fruit.

» Its diet of fruit stem ends is responsible for substantial fruit loss.

» There's a halt to the tree's expansion.

» The flowers and fruit drop off too soon.

Pest identification

» **Eggs:** Eggs are deposited in clumps and covered by a cottony substance for protection.

» **Nymphs:** Amber-colord nymphs with a white waxy coating and a few stray filaments.

» **Adult:** Winged, with a large antenna, and no mouth parts, the adult male flies. Females lack wings, have a flatter body shape, and have short, waxy filaments along the edges of their bodies.

Management

» Assemble the ruined foliage and throw it away.

» Make use of sticky traps of the 5-centimeter variety on fruiting stalks.

» Spray Water diluted with 1.5 ml of dimethoate and 2.5 ml of kerosene, 1 g of carbaryl and 2 ml of malathion, or 2 ml of chlorpyriphos.

» *Leptomastix dactylopii,* an exotic parasitoid, will be released at a density of 5,000 adults per hectare (Need based under expert supervision)

» The Field Discharge After blooming, a population of 10 predatory beetles per afflicted tree of *Cryptolaemus montrouzieri* (starved overnight) is considered normal.

sylla, *Diaphorna citri*

Damage symptoms

» Adults and nymphs alike feed on the cell sap of infested plants, resulting in symptoms like leaf curling, defoliation, and profuse flower drop, which has a devastating impact on fruit set and can even result in the complete death of branches, known as dieback.

» It is thought that in addition to its saliva, it secretes poisons that cause the death of non-attacked branches.

» Fruit prices are affected by sooty mould, which is spread by nymphs, which exude a crystalline honeydew that coats the shoots and leaves on which it grows.

» Because of their small size, lack of flavorful juice, and lacklustre flavour, the fruits are a quality and quantity disaster.

» Psylla is most active in the spring and monsoon seasons and has a lower frequency of activity in the winter.

» The affected sections of plants wither and perish.

» The nymphs in their fifth instar are responsible for spreading the greening bacteria, hastening the insect's demise.

Pest identification

» **Nymphs:** orange, oblong, and squat in appearance.

» **Adult:** Adults are tiny insects with wings that extend past the tip of the abdomen and a lustrous black body dusted with grey.

Management

» It's time to cut back on those dried out, diseased shoots.

» It's not a good idea to plant citrus trees near a host plant like curry leaf (Murraya koenigi).

» Using a 0.025% foliar spray of Phosphamidon, 0.025% of Quinalphos, and 0.025% of Thiometon.

» Up to 95% parasitism can be attributed to the eulopid nymphal parasitoid Tamarixia radiata.

» Nymphal psylla are preyed upon by predators such the Mallada boninensis, Apertochrysa crassinervis, and Brumus suturalis.

Whitefly, *Dialeurodes citri*

Damage symptoms

» The months of March–April, July–August, and October–November corresponds to the onset of new flushes, and hence are the most common times of observation.

» In Punjab, kinnow and sweet orange are taken quite seriously.

» They weaken plants by removing vital cell sap with their sucking mouth parts, a problem that affects both nymphs and adults.

» Consequently, the grass and trees lose their vibrant green colour, turn brown and curly, and eventually fall to the ground.

» The honey dew that nymphs secret on leaves attracts black sooty mould, which stunts photosynthesis.

» Trees that are heavily infested with the pest become sick and frail, only able to yield meagre harvests of tasteless fruit.

» The tree's ability to bear fruit and the quality of its fruit are also drastically altered.

» Blackened fruits taste bad and don't sell well because of it.

Pest identification

» **Nymph:** an oval, flat thing that is attached to the underside of a leaf in a very secure fashion.

After the initial moult, it cannot be changed.

» **Adult:** Mealy-white adult insects have four wings that don't spread more than 3.2 mm. The wings of both sexes are covered in a powdered white wax.

Management

» Use a spray solution of 1 ml of Monocrotophos, 1.25 g of Acephate, and 1.5 ml of Phosalone per L of water, alternating between these three solutions every 15 days during the flushing season.

» The aphid parasitoids *Encarsia lahorensis* and *E. bennetti* are effective.

» Whiteflies are controlled by the parasitic fungus *Cladosporium sp.* and *Ascheronia aleyrodis.*

Fruit sucking moth, *Otheris materna, O. ancilla, O. fullonica*

Damage symptoms

» Heavy fruit drop in mandarins and sweet oranges, especially in September and October, is caused by fruit sucking moth, a common and harmful pest of maturing fruits.

» The lovely moths are strong and agile, making them tough to contain.

» The adults eat the fruit by puncturing it to get at the juicy pulp within, which eventually causes the fruit to decay and fall to the ground.

» Due to rot caused by fungal and bacterial infections introduced by puncture a large percentage of these fruits are lost before they ripen (40%).

» The adult moths fly between August and December, with the most activit occurring in the fall months of September and October.

Pest identification

» **Larvae:** velvety dark mottled all over with orange, blue, and yellow patches

» **Adult:** The mature moths are easily recognisable by their bulky buil and bright orange wings.

» *Otheris materna*: three black patches on the front wing of an *Otheris materna*

Management

» Wipe off the *Tinospora cardifolia* and *Coccules pendules* that are feeding o the weeds.

» Attract and kill moths with a light trap or a food lure.

» For two hours each evening, smoke should be generated in the orchard t deter adult moths.

» The fruit should be individually wrapped in polythene bags (500 gauge) o palmyrah baskets.

» Cultivating tomatoes as a trap crop in fruit orchards can help catch th adult moths that can then be removed.

» To poison bait, mix 2 litres of water with 20 gm of Malathion WP or 5 ml of Diazinon plus 200 gm of jaggery and a few tablespoons of vinega or fruit juice.

Bark eating caterpillar, *Indarbela quadrinotata, I. tetraonis*

Damage symptoms

» Infestations happen in orchards that have been poorly maintained and d not practise good hygiene.

» Even on the newest Nagpur mandarin orchards, it is spreading at an alarmin rate.

» Seventeen tunnels were found in a single 8-year-old tree.

» These kinds of orchards appear like they're dying off.

» At night, grubs feed on the bark that covers the tree's exposed wood burrowing into the forked trunk and the branches near the joints. This reduces the tree's productivity by blocking the sap flow.

» This is where they go to hide during the day: the tunnel.

» Outer tunnels may be adorned with distinctive webbings made of wood frass and grub faeces, which are a telltale sign of a pest infestation.

» This sort of infestation is typical in orchards that have been around for a while but have been ignored.

» An attack by a bark-eating caterpillar drastically reduces a tree's vitality and shortens its lifespan.

Management

» To eliminate the insect, a cotton wad soaked in either 10 ml of gasoline or kerosene per tunnel, or 0.1% dichlorvos, or 0.02% monocrotophos, is inserted into the tunnels and then sealed up with mud.

» Within 10 minutes of injection with a plastic injection syringe, 4–5 ml of insecticidal solution (4 ml Dichlorvos or 2 ml Monocrotophos diluted in one litre of water) can kill any concealing larvae.

» After removing the frass and faeces pellets, it is planned to spray the affected area with the aforementioned insecticides in an effort to maintain pest control.

Trunk borer/stem borer, *Monohammus versteegi, Chelidonium cinctum*

Damage symptoms

» Following its hatching, the grub begins a horizontal bore in the sapwood near the ground, and then it begins a vertical tunnelling upward along a winding path.

» The grub and the pupa both spend their time inside the hole.

» Next summer, after spending the winter in the pupal chamber, the adult will emerge through an exit hole that is 1.5 to 2.0 feet above the original hole.

» It is possible to identify citrus trunk borer by the appearance of resinous exudation and sawdust-like powder on tree trunks at ground level.

» The midribs of a leaf are the borer's preferred meal, therefore the edges are safe from damage.

» The citrus fruit crop suffers as the affected branches eventually dry out.

» At heights of up to 2.5 metres from the ground, eggs are placed on the trunks and primary branches of trees.

Management

» Please keep the sink spotless.

» Check the trunks frequently for signs of damage or swelling.

» When spotted, capture and eliminate immediately.

» To eliminate the larvae, metal wire should be inserted into holes.

» Put a clump of cotton dipped in kerosene or gasoline (enough to absorb about 10 ml) into the mouth, then seal it off with a mud plug.

» 5 ml of Carbaryl (2 ml/L of water), DDVP (1 ml/L of water), or Monocrotophos (1 ml/L of water) should be swabbed into the stem tunnel up to a height of 2 metres.

» Trunk borer is no match for the white muscardine fungus, *Beauveria bassiana.*

» Take down and burn any trees that are diseased or aren't producing anything.

Thrips, *Scirtothrips aurantii*

Damage symptoms

» Adults and juveniles pierce leaves to get sap, and both feed on fruits.

» The inward curling of leaves.

» The fruit has a halo shape.

» Skin has irregular, mottled areas.

» The citrus thrips makes tiny puncture marks in the skin of fruit, which eventually turn grey or silvery.

» Since they are larger than first instars, second instar larvae cause the most harm by feeding behind the sepals of young fruit.

» As fruits mature, scar tissue from damaged rind tissue radiates outward from the sepals.

» After the petals fall off, the fruit is at its most vulnerable to scarring until it reaches a diameter of around 3.7 cm.

» Fruit on the outside canopy is more at risk from thrips, as well as wind and sun damage.

est identification

» **Eggs:** The banana-shaped eggs measure at just 0.2 mm in length.

» **Larvae:** Light orange-yellow to white in colour, larvae resemble adults but lack wings.

» **Adult:** The wings of an adult are a fringed yellow. The average length of a female is 0.6 to 0.9 mm (0.02 to 0.04 in). Similar to males, but smaller in stature. The third through eighth segments of the antennae are grey.

Management

» The broken pieces of plants should be gathered and thrown away.

» In order to control the thrips problem in greenhouses, gentle insecticides such soaps, botanicals like Sabadiella sugar, and even just washing with clean water should be used.

» At the peak of growth, spray systemic insecticides like Malathion (0.05%) or Carbaryl (0.1%).

» Increase the presence and activity of natural predators like syrphids and chrysopids.

» When used in conjunction with predatory mites, Ryania and Sabadiella play a more significant role in thrips management.

Cottony cushion scale, *Icerya purchasi*

Damage symptoms

» Adults and juveniles alike feed on leaf sap.

» Infest older leaves frequently.

» Can lead to leaf yellowing and dropping.

» When scales infest a plant heavily, it can be fatal to the young shoots.

» Infestations in fruit are rare but do happen.

» Honeydew emitted by cottony cushion scale coatings fruit and encourages the formation of sooty mould fungus, which hinders photosynthesis, weakens the plant, and renders fruit unappealing.

est identification

» Eggs: red and rectangular, with a cottony ovisac.

» **Adult female:** Waxy reddish-brown to yellow skin covers the adult females' body.

» **Adult male:** Dark red body, dark antennae, and dark legs characterise the adult male. A black, diamond-shaped patch sits atop the dorsal region the insect's trunk (thorax).

Management

» Preparing for spring by spraying dormant oil in late winter.

» The oil used in horticulture can be sprayed at any time of the year.

» Put an end to the ant infestation that could offer scale a competitive edge

» Infestations of cottony cushion scale can be effectively suppressed by releasing 10 predatory vedalia beetles per afflicted tree immediately upon discovering the presence of adults.

» Introducing Australian ladybugs to the great outdoors.

Mites, *Eutetranychus orientalis*

Damage symptoms

» The citrus mite not only eats away at the foliage, but also the fruit itself leaving behind unsightly brown spots on the peel and lowering the fruit already poor market value.

» When the insect eats the upper surface of the old leaves, it creates a mottle appearance and makes the leaves look dusty.

» The citrus mite feeds on the juices of citrus plants.

» When the infestation is really bad, the stripping will spread and eventually become dry necrotic regions.

» Typically, leaves fall off and twig dieback begins.

» When mature trees begin to sprout new growth, it provides the ideal environment for the pest to rapidly and continuously multiply, eventually causing an epidemic.

Management

» The highest mite mortality rate was seen with 0.025% Tetradifon, followed by 87.57% with 0.025% Phosphamidon.

» An alternative is to apply a foliar spray of either Dicofol (1.5 ml), Monocrotophos (1.0 ml), Oxydemeton methyl (1.5 ml), or Wettable sulphur (3 g/L of water).

» It is especially important to water the trees thoroughly in the late summer, when they are under the most strain. Pests like the pirate bug (*Orius* spp.) and the *Euseius tularensis* feast on citrus mites.

DISEASES

Foot rot/gummosis, *Phytophthora citrophthora*

Symptoms

» Leaves turn yellow, then the bark cracks and there is excessive gumming on the surface; these are the signs.

» When gumming gets bad enough, it rots the bark and causes a girdling effect that eventually kills the tree.

» The dry, shrinking, cracking, and shredding of bark in such areas occurs in long, vertical bands.

» There's an infection that's spread to the crown roots.

» The plant typically produces abundant blossoms right before it dies, but then it withers and falls over before its fruits have a chance to develop.

Favorable conditions

» Soil that is heavy, wet, acidic (pH between 5.4 and 6.5), and warm (25 to 28°C) is ideal.

Survival and spread

» Mold can live on dead fruit, branches, leaves, and in tree crevices.

» Rainwater, irrigation water, wind, and insects can all help disseminate sporangia.

Management

» Drainage needs to be improved.

» Do not over-irrigate.

» During cultivation, take care not to damage the crown roots or the base of the stem.

» Remove the damaged skin and then cover it with Bordeaux paste.

» From July through September, once a month, soak the soil surrounding the tree with 20 litres of 1% Bordeaux mixture/tree, covering an area 1 m in diameter.

» Every year before the rainy season begins, healthy trees should have Bordeaux paste applied to their trunks up to a height of 50 to 75 cm above ground level.

» One % Bordeaux mixture sprayed alone or in combination with tin sulphate 0.3% Difolatan, 0.2% metalaxyl-mz, or 0.1% fosetyl-al (Aliette).

» Soil should be amended with FYM containing 2 kg of *Trichoderma* spp.

» *Glomus fasciculatum* mycorrhiza should be applied to the soil.

» If the lesion is smaller than half the diameter of the stem, you can treat it with Bordeaux paste after cutting off the infected bark and leaving 1.2 cm of healthy bark.

Powdery mildew, *Acrosporium tingitanium*

Symptoms

» Infection typically manifests itself initially on the fresh flush or young growth.

» The upper leaf surface is where the white, 'powdery,' spores typically form.

» New leaves are a drab grey-green at first.

» Mildewy leaves often curl upward and twist at the tips.

» Dieback of twigs and branches is a common problem among plants.

» As a result of a severe infection, leaves fall off.

» Young fruit can also grow white, 'powdery,' spores.

» Precocious decay of infected fruit.

» Lower production.

Favorable conditions

» Morning rain and limited sunshine.

» Occur between the months of October through March particularly at higher altitudes.

» Mandarin, lemon, and sweet orange all carry a serious tone.

Survival and spread

» The disease spreads via microscopic, powdery spores that can hang around for a while on dead leaves.

» The wind, humans (clothes, hands), equipment (e.g., pruning tools, mechanical harvesters, or hedgers), and vehicles can all spread the spores far and wide.

» One major risk is the spread of diseased citrus seedlings through the transportation of infected planting material.

Management

» Pruning water shoots on a regular basis helps to prevent the spread of disease.

» It's important to remove and dispose of any diseased plant pieces.

» At the first sign of attack, spraying with Tridemorph, Triadimefon, Dinocap, and Benzimidazole fungicides should be performed, and subsequent applications should be spaced out by 10 days for optimal disease management.

» In addition, sulphur dusting once every eight days is indicated for disease prevention.

Scab/Verucosis, *Elsinoe fawcetti*

Symptoms

» The earliest symptoms of the disease appear as tiny spots of pale orange on the leaves.

» The lesion causes the leaf tissue to become deformed into stiff, hollow, conical growths.

» The underside of leaves usually suffer from lesions first. They can be seen on both sides of the leaf once they've broken through.

» Infected regions merge, spreading across a wide area.

» Sometimes the leaves will become wrinkly, stunted, or misshapen.

» Similar sores develop on twigs.

» Lesions on fruit manifest as corky protrusions that eventually scab off.

» Young fruits have cream-colored scabs, whereas ripe fruits have dark olive-grey scabs.

» Fruits that are assaulted while still developing take on an abnormal appearance, with conspicuous warty projections.

» It's a case of premature extinction.

Favorable conditions

» Thrive in damp, cold conditions.

» They are especially vulnerable while the leaves are young.

Survival and spread

» The pathogen spreads through conidia and overwinters as ascospores.

Management

» Collect and dispose of the infected leaves, branches, and produce.

» When new shoots sprout in the spring, it's best not to water them from above.

» Trees need to be pruned on a regular basis to keep them healthy, open, and clear of deadwood. This improves airflow and spray coverage within the tree.

» The spraying of 0.1% Carbendazim is quite effective.

» Hexaconozole (0.01%) and streptomycin sulphate (100 ppm) were sprayed twice, once a month apart, beginning one month after fruit set and again a month later.

» Citrus fruits such as grapefruit, oranges, and limes are known for their resilience.

Anthracnose, *Colletotrichum gloeosporoides*

Symptoms

» Leaves being discoloured, branches withering, leaves falling off too soon, fruits becoming discoloured after harvesting, and so on.

» Leaves and branches that are dying are covered in dark fungal spores, which are how the disease is disseminated.

» Valencia and navel oranges, grapefruit, and even lemons can develop anthracnose spots on their rinds as they ripen.

» Fruit on stressed trees with old, dead wood is especially vulnerable to the disease.

Survival and spread

» Mold and fungus can live on wood that has naturally decomposed.

» In short distances, it can be dispersed by splashing rain, heavy dew, or overhead irrigation, all of which disperse spores into the air.

Management

» Irrigation, fertilisation, cultural practises, and pesticides can all help reduce the severity of the disease.

- » Dropping fruit can be minimised by removing dead branches in the winter and spraying with a 1% Bordeaux mixture in February, March, and September.
- » Protect the cut ends by applying Bordeaux paste.
- » After pruning, trees should be sprayed with 0.1% Bavistin or 0.2% Captafol three times.
- » The best outcomes were seen when using Prochloraz or Benomyl to prevent the disease from spreading in Coorg mandarins while they were being stored.
- » Spray 40-100 L/tree of zinc sulphate + copper sulphate + hydrated lime or 350-450 ml of Azoxystrobin (Abound) 2F per tree.

acterial canker, *Xanthomonas compestris* pv. *citri*

ymptoms

- » Citrus fruits like lemon, lime, and grapefruit that are high in acid are impacted. Extremely uncommon on tangy citrus fruits.
- » Leaves, young branches, elder branches, thorns, and fruits are all susceptible to disease.
- » Small, watery, translucent spots called lesions first emerge on the underside of leaves and eventually spread to the upper surface.
- » Spots turn white or grey and take on a crater-like, corky look.
- » Usually, lesions are spherical, develop on both sides of a leaf, and are encircled by a halo of yellow.
- » Twigs develop lengthy lesions, girdle, and eventually perish.
- » Comparable to leaf cankers, but more apparent, cankers on fruits have a central depression that looks like a crater.
- » Usually, lesions on fruit only spread to the rind, but it can also lead to skin damage like cracks and fissures.
- » Canker lesions lower an apple's resale value.

avorable conditions

- » 20 to 30 °C.

urvival and spread

- » Holds out for six months in contaminated leaves.

» Blowing and pouring rain creates a soaking mess.

» Leaf miner damage aids the bacterium's ability to invade the plant.

Management

» Inspect for leaf miner infestations during times of new growth.

» Remove diseased branches before the monsoon season begins.

» The disease's severity in the groves was greatly diminished by the additic of windbreaks, and the other methods of disease control were also great bolstered.

» Streptomycin sulphate (500-1000 ppm), phytomycin (2,500 ppm), or copp oxychloride (0.2%) sprays every two weeks.

» Neem cake solution applied to plants can help prevent canker.

» *Bacillus subtilis* and *Aspergillus terreus,* both of which were isolated fro diseased citrus leaves, proved efficient when applied as phylloplar antagonists.

» The canker resistance of acid lime cv. 'Rasraj'. Sweet oranges (such as tl pre-immunized Pera, Navelina, Valencia, and Folha Murcha), mandarii (such as the Ponkan, Dancy, and Mexerica do Rio), and Tahiti limes a typically suggested for planting due to their canker resistance.

Greening, *Liberobactor asiaticum* (Phloem limited bacteria - like organism)

Symptoms

» When a leaf matures, the tissue directly close to the midvein always turi a consistent yellow, even in the first flush.

» The interveinal tissue and the main lateral veins of a leaf often turn yello together.

» Die-back signs can be seen on the secondary growth, which is erect, sho: has little leaves, and numerous buds.

» A zinc or iron shortage might be seen in the leaf symptoms.

» Later, whole lamina including veins becomes chlorotic with a few da: green islands.

» Leaf stunting, scant foliation, twig die back, poor harvest of mostly gree useless fruits.

» Small and asymmetrical with a curled columella. The side that faces the st turns a vibrant orange, while the opposite side stays a muted olive green.

» High in acidity and low in juice and soluble solids. Not fit for consumption or further use in any way. undeveloped, black, and abortive seeds

Favorable conditions

» There is a plethora of vectors present.

Spread

» Symptomatic branch wood.

Diaphorina citri, a vector psyllid.

Management

» Nursery stock should be propagated using bud wood that has been certified to be free of pathogens.

» A temperature of 47 °C can be maintained for the bud wood for up to four hours.

» Before budding, the bud sticks could be dipped in 500 ppm Ledermycin or Pencillin extract together with 500 ppm Bavistin, effectively wiping out the greening pathogen.

» Spraying with 0.05% Monocrotophos or 0.15% Metasysttox at regular intervals will keep the psyllids under control.

» Fortnightly spraying with 500 ppm Streptocycline decreases disease incidence by stopping the pathogen from reproducing.

Tristeza or quick decline, *Citrus Tristeza Virus* (CTV)

Symptoms

» Fine pitting, or honeycombing, can be visible on the inner bark face just below the bud union in the vulnerable rootstock region of the tree.

» Sometimes, when held up to the light, you can see the veins in immature leaves of acid lime plants clearing out.

» When the bud union is damaged, necrosis of the phloem sieve occurs, causing the roots to starve.

» Without their feeder roots to provide water to the canopy, plants quickly wilt and die.

» Trees that have been infected with the disease struggle under the weight of their loads.

» Seedlings and bud-lings of limes grown from any root source are susceptible to the disease.

» To the contrary, grapefruit and sour orange rootstocks are vulnerable for mandarin and sweet orange seedlings.

» Insufficient nutrients cause the leaves to abscise.

» The symptoms include root rot and die-back of the branches.

» Reduced fruit production leaves nothing but a shell. The fruits are tiny.

» Tree died or became stunted. As a result, yield drops drastically.

Spread

» The application of diseased bud wood.

» The *Toxoptera citricidus* aphid plays a crucial role as a vector.

Management

» Crucial is the development of methods for mass-producing virus-free trees such as shoot-tip grafting and thermal treatment.

» Disease-resistant fruit trees, sow disease-free seedlings in Rough lemon or Cleopatra mandarin rootstocks.

» Put to good use virus-free, true-to-type seedlings from a nucellar.

» Citrus trees for a nursery should be grown in a net house to protect them from pests.

» Use seedlings that have been immunised against a moderate strain of tristeza to grow acid lime.

» In order to prevent the further development of tristeza in the orchard, periodic spraying with pesticides such as 0.05% Monocrotophos or 0.15% Metasystox is recommended.

References

Anjos, D. V., Tena, A., Torezan-Silingardi, H. M., Pekas, A., & Janssen, A. (2021). Ants affect citrus pests and their natural enemies in contrasting ways. Biological Control, 158, 104611.

Folimonova, S. Y., and Sun Y. D. (2022). Citrus tristeza virus: From pathogen to panacea. Annual Review of Virology, 9, 417-435.

Ghosh, D. K., Motghare, M. and Gowda, S. (2018). Citrus greening: overview of the most severe disease of citrus. Adv Agric Res Technol J, 2(1), 83-100.

Jivrag, K. P., Patil, B. V., Pavhane, S. G., Savalkar, J. L., & Wadhekar, N. B. (2022). Seasonal incidence of citrus blackfly, citrus whitefly and their natural enemy during Mrig bahar. The Pharma Innovation Journal, 11(11), 1825-1828

Khanchouch, K., Pane, A., Chriki, A. and Cacciola S. O. (2017). Major and emerging fungal diseases of citrus in the Mediterranean Region. Citrus Pathology, 1(10.5772), 66943.

Setamou, M., Soto, Y. L., Tachin, M, and Alabi, O. J. (2023). Report on the first detection of Asian citrus psyllid Diaphorina citri Kuwayama (Hemiptera: Liviidae) in the Republic of Benin, West Africa. Scientific Reports, 13(1), 801.

Timmer, L. W., Garnsey, S. M, and Broadbent P. (2003). Diseases of citrus. Diseases of tropical fruit crops, 163-195.

Urbaneja, A., Grout, T. G., Gravena, S., Wu, F., Cen, Y., & Stansly, P. A. (2020). Citrus pests in a global world. In The Genus Citrus (pp. 333-348). Woodhead Publishing.

8 Custard Apple

INSECT PESTS

Mealy bug, *Maconellicoccus hirsutus, Ferrisia virgata*

Damage symptoms

- » They feed on plant parts (including fruits and flowers) and sap.
- » The afflicted plant tissues shrink and retain their abnormal shapes.
- » Roots that have been infected form a spongy tissue.
- » » The honey dew that mealy bugs excrete feeds sooty mould on plants and fruits and attracts black ants, both of which aid in the spread of mealy bugs
- » Both adults and invertebrates can be found resting on the leaves, stems, and fruits (between segments).
- » Observable leaf yellowing.
- » Reduced fruit size means they no longer sell at a premium.

Pest identification

- » **Nymph:** Nymphs can be anything from lemon-to-cream coloured.
- » **Adult:** Females reach maturity as apterous, elongate insects with white waxy secretions covering their bodies.

Management

- » Remove and discard any infected branches or fruit.
- » Two applications of a spray containing either 0.05% dichlorvos, 0.04% diazinon, 0.05% monocrotophos, or 0.05% chlorpyriphos, once during the

period of new growth and shoot formation and again during fruit set, are recommended.

» Mealy bug predators include *Spalgis epius, Scymnus coccivora, Cryptolaemus montrouzieri, Triommata coccidivora,* and *Cacoxenus perspicaux.* Release 10-per-tree infestation of *Cryptolaemus montrouzieri* insects

» *M. hirsutus* and *F. virgata* were found to be parasitized by the encyrtid parasitoids *Anagyrus dactylopii* and *Aenasius advena.*

» The mortality rates of mealy bugs ranged from 89.31 % when treated with *Verticillium lecanii* at 2.0 g/liter of water + Ranipal 1 ml/liter of water, to 86.95 % when treated with V. lecanii at 2.0 g/liter of water + Teepol 1 ml/liter of water, to 85.81 % when treated with *V. lecanii* at 2.0 g/liter of water alone.

Queensland fruit fly, *Bactrocera zonata*

Damage symptoms

» They lay their eggs (a'sting') in fruit, and the larvae feed on human flesh.

» Damaged fruit is easy to spot because rot sets in quickly and the skin around sting scars turns a different colour.

» The almost-ripe fruits were ravaged by maggots.

» Consume just internal fruits.

» In almost ripe fruit, softening occurs around the egg-laying sites, and this is occasionally accompanied by yellowing and exudate.

» The affected fruits get shrivelled and deformed, and eventually perish and fall off.

Pest identification

» **Larva:** The larvae are apodous maggots that are yellow in colour.

» **Adult:** The wings of an adult are clear and the body is a pale brown.

Favorable conditions

» Mid- to late-summer is typically the worst time of year for damage caused by the Queensland fruit fly.

Survival and spread

» In order to stay alive, fruit flies require a consistently damp environment.

» Strong winds can also carry away Queensland fruit flies, which have a range of up to 1 km on their own.

Management

- » It's best to pick fruits just before they reach full ripeness.
- » Recycle the rotten fruit by collecting it from the ground and burying it a pit.
- » Pupae can be uncovered during summer ploughing.
- » Fruits can be soaked in hot water (45 to 47 °C) for 60 minutes to kill a eggs or maggots that may be present.
- » Use Methyl eugenol sex lure traps to observe fly activities.
- » Insecticides (Fenthion 100 EC 1 ml/L, Malathion 50 EC 2 ml/L, Dimethoa 30 EC 1 ml/L, or Carbaryl 50 WP 4 g/L) and sugar or jaggery (10 g/I can be used to create a bait spray.
- » Apply sprays twice, with a 2-week interval between each application, ju before the fruits reach full ripeness.
- » Apply a solution containing 25 g of malathion per hectare and 1 gramm of methyl eugenol per hectare as bait.
- » The spray concentration of 1ml/L of Malathion 50 EC is 0.1%.
- » Dispersal of Predators and Prey Both the *opium poppy* and the *Philippi spalangia* are used as a form of compensation.
- » Put out the Encyrtid parasitoids, such as *Anagyrus dactylopii* and *Aenasi advena.*
- » Set predators free, *S. epius, S. coccivorus, C. montrouzieri, T. coccidivora,* an *C. perspicaux.*

Fruit borer, *Heterographis bengalella*

Damage symptoms

- » Caterpillars eat the pulp inside of fruits, damaging the mesocarp in th process.
- » Caterpillar droppings can be seen close to the fruit's entry points.
- » The development of infected fruits stops, and the fruit eventually falls of

Pest identification

- » **Larva:** Adult caterpillars of the gray-to-black species reach a length of 20 mn
- » **Adult:** The transparent wings of an adult brown butterfly.

Management

» Recover the infected fruit and throw it away.

» Start selling fruits in bags.

» Malathion 0.1% should be sprayed twice, once during blooming and again during fruit setting.

» When spraying, use 2 ml of Quinalphos, 2 ml of Carbaryl, and 2 ml of Chlorpyriphos per litre of water.

» Parasitism of larvae using braconid insects (*Apanteles* spp.)

» Ground beetles, rove beetles, *Chrysoperla zastrowii sillimi*, coccinellids, reduviid bugs, preying mantids, big-eyed bugs (*Geocoris* sp.), pentatomid bugs *(Eocanthecona furcellata)*, and so on are all predators.

Seed borer, *Bephratelloides cubensis*

Damage symptoms

» Insect larvae develop inside the seeds as they consume the endosperm.

» Adult wasps emerge from their burrows on the outside of fruit, where they frequently cause fungus infection and the fruit to shrivel and rot.

Management

» If the fruit is less than 5 cm in diameter, placing it in a bag to keep it clean is a surefire way to ensure that it will be eaten without any contamination.

» Use insecticides such as pyrethrin, spinosad, sulphur, azadirachtin, *Bacillus thuringiensis*, buprofezin, imidacloprid, insecticidal soap, methidathion, pyrethrin, or sulphur.

Fruit-spotting bug, *Amblypelta nitida*

Damage symptoms

» Adults and juveniles alike pierce fruit and drain its juices for sustenance.

» When feeding, they plunge their lengthy beaks into the fruit and expel saliva containing enzymes that digest the green fruit cells.

» As a result, extensive tissue disintegration occurs at a subcellular level.

» Damage appears as tiny black dots on the shoulders of developing fruit, measuring between 2 and 10 mm in diameter.

» About 1 cm of the fruit's flesh has been harmed.

» Roots, stems, and even young shoots can be harmed.

» Fruit is constantly pierced by adults and nymphs, and the young fruit fall off after 5-10 days.

» While older fruit might not fall off, its quality definitely declines.

Favorable conditions

» The hottest months (>32°C) are the worst for fruit spotting bugs.

Management

» If there is a lot of scrub, don't put trees nearby.

» Predators on fruit flies include assassin bugs and spiders.

» The pests can be effectively managed with the use of methidathion (such as Supracide). If damage persists after two to three weeks, apply the spray again. Even after two sprayings, you should keep an eye out for falling fruit and reapply the treatment if necessary.

DISEASES

Fruit spot, *Pseudocercospora purpurea*

Symptoms

» Diffuse spots, which initially manifest as indentations on the fruit, are an early indicator of the disease.

» Small (up to 15 mm) and dark purple to grey spots.

» Many spots will join together to form larger, unsightly patches.

» As the condition worsens, large necrotic patches become increasingly common.

» In advanced stages, the skin develops superficial fractures and becomes so hard that it hampers the normal growth of fruit.

Management

» Protective spraying with a mixture of Mancozeb, pyraclostrobin (Cabrio) and paraffin oil should be performed regularly (Biopest).

» Biopest, a novel paraffin oil composition, is effective both on its own and in combination with Mancozeb.

» Apply oil and Copper fungicide on the cut stumps and leaf litter of the pruned trees.

» Increase spray penetration by opening up the trees during pruning. Clean up the orchard of any trimmings.

» Tree flanges should be trimmed to a height of 50 cm from the floor. This will ensure the best possible spray coverage and reduce humidity levels in the canopy.

» Check on fruit for signs of infection on a regular basis throughout the season so that spraying can begin before diseases become severe.

Anthracnose, *Colletotrichum gloeosporioides*

Symptoms

» Dark brown to black patches, similar to black canker, can be found on immature fruit.

» In a second manifestation, the same fungus causes the fruit to get covered in particles that look like pepper.

» Poorly impacted fruit will have a conglomeration of these pepper patches.

» Occasionally, leaf symptoms may be observed as well.

Favorable conditions

» Climates that are wet and windy are ideal for the spread of the disease.

Management

» Copper oxychloride fungicide and paraffinic oil should be sprayed on dormant trees twice (Biopest).

» Fungicides such Mancozeb or Pyraclostrobin (Cabrio) or Paraffinic oil can be sprayed every two to three weeks (Biopest).

» Only apply Prochloraz (Octave) pesticide spray during the blossoming and early fruit set stages.

Bacterial wilt, *Ralstonia solanacearum*

Symptoms

» Yellow or white in colour, most leaves are a mild shade of green.

» Dye staining of the wood beneath the bark is seen on the lower trunk.

» Decomposition of bark around the crown occurs at or near ground level.

» A black staining of water-conducting tissue can be seen through a piece of bark cut from above the affected area.

» There may be a quick wilting and death of young trees, often accompanied by significant defoliation.

» The remaining leaves are a drab, green colour and hang nearly vertically from the tree.

» Older trees typically deteriorate gradually over the course of two years, with minimal leaf yellowing if at all.

» A dark staining can be seen in the water-conducting tissues of the basal trunk and big roots of affected trees.

» This is especially common in trees that have recently begun bearing fruit.

Spread

» This bacteria survives in crop wastes and weed hosts and spreads across the soil. It can travel through water, especially if it's flowing downhill, and it can perhaps spread by root-to-root contact.

Management

» The lives of afflicted trees may be extended by mulching and lowering agricultural loads.

» Try not to plant in spots where you last grew tomatoes, potatoes, eggplant or capsicum within the prior two years.

» Avoid planting in low spots and raise the ground to create mounds to let water run off.

» Put to use the disease-proof cherimoya rootstocks.

References

Al-Ghoul, M. M., Abueleiwa, M. H., Harara, F. E., Okasha, S., & Abu-Naser, S. S. (2022). Knowledge Based System for Diagnosing Custard Apple Diseases and Treatment.

Butani, D. K. (1976). Insect pests of fruit crops and their control-custard apple. Pesticides, 10(5), 27-28.

Gargade, A., & Khandekar, S. (2021). Custard apple leaf parameter analysis, leaf diseases, and nutritional deficiencies detection using machine learning. In

Advances in Signal and Data Processing: Select Proceedings of ICSDP 2019 (pp. 57-74). Springer Singapore.

Gargade, A., & Khandekar, S. A. (2019, March). A review: custard apple leaf parameter analysis and leaf disease detection using digital image processing. In 2019 3rd International Conference on Computing Methodologies and Communication (ICCMC) (pp. 267-271). IEEE.

Khaliq, G., Ullah, M., Memon, S. A., Ali, A., & Rashid, M. (2021). Exogenous nitric oxide reduces postharvest anthracnose disease and maintains quality of custard apple (Annona squamosa L.) fruit during ripening. Journal of Food Measurement and Characterization, 15, 707-716.

Mayers, P. E., & Hutton, D. G. (1987). Bacterial wilt, a new disease of custard apple: symptoms and etiology. Annals of applied biology, 111(1), 135-141.

Sharma, H. C., Sankaram, A. V. B., & Nwanze, K. F. (1999). Utilization of natural pesticides derived from neem and custard apple in integrated pest management. Azadirachta indica A. Juss.., 199-213.

Shukla, R. P., & Tandon, P. L. (1984). India-insect pests on custard apple. Plant Protection Bulletin, FAO, 32(1).

Shylesha, A. N., & Mani, M. (2022). Pests and Their Management in Custard Apple. Trends in Horticultural Entomology, 803-816.

9 Fig

INSECT PESTS

Stem borer, *Bactrocera rufomaculata*

Damage symptoms

- » Sapwood galleries caused by wood boring insects.
- » Grubs cause harm by chewing a zigzag path through the bark and inn wood (xylem), leaving behind frass made of cellulose and excrement.
- » To create their crooked tunnels, grubs eat through the sapwood.
- » Taking blood and eating the veins and arteries.
- » Denial of essential nutrients and water to the affected tissue.
- » First-stage terminal shoot drying out.
- » Sap and frass both occasionally seep through the cracks.
- » When a strong wind blows, the attacked branches of a tree are weakene and eventually break off.
- » The tree's ability to produce fruit is severely stunted.

Pest identification

- » **Egg:** The egg is a 6.2 mm long, brownish-white cylinder with tapered, rounde ends.
- » **Larva:** Adult grubs (larvae) can grow to be up to 10 cm in length; they a cream in colour with a dark brown head.
- » **Adult:** They reach a length of 3 to 5 cm as adults and have a fine, grey textu and a black coloration.

Management

» Remove diseased dry branches and dispose of them in a fire.

» Fill holes with mud after first filling them with a cotton wad that has been soaked in kerosene, petrol, or Dichlorvos.

» The hole can be filled after being injected with a 3:1 mixture of ethylene dichloride and carbon tetrachloride.

» Spraying the fig tree stem with 0.1% Chlorpyriphos or using coal tar coated paper will deter oviposition.

Scale insect, *Parlatoria olcae, Aspidiotus cydonige, Pseudoccoccus lilacinus*

Damage symptoms

» The nymphs do their damage by inserting their stylets into the plant while feeding, which can result in the premature loss of leaves and the death of branches.

» Large volumes of honeydew secreted by these scales may promote the formation of sooty mould.

» White waxy mounds form on the stems and undersides of the leaves as well. The dark sap-sucking insects' egg masses look like this.

» There may be some die-back of shoots or branches if the infestation is bad enough.

» Plants can be weakened by these insect pests if they are draining the life force out of them.

» Having a lot of people around can be fatal to the host.

Pest identification

» **Egg:** Eggs are orange at first, then become pink just before they hatch. Pre-hatching inspection of the eggs reveals a pair of red eyespots.

» **Crawlers:** They are bright pink, orange with red or black eyes, 0.2 - 0.3 mm long. First instars are 0.6 - 0.8 mm length and 0.2 - 0.4 mm broad.

» **Adult:** Adult A hallmark of black and white art is the sharpness of the letter H's ridges. The size of adults is vary, 1.9-5 mm length, 1-4 mm wide and 1.2-2.5 mm high.

Management

- » *Metaphycus helvolus,* a parasitic wasp, feeds exclusively on scale insects.
- » Deltamethrin spray (contact insecticide).
- » Plant oils to spray.
- » Some parasitoids are the nematode *Encarsia perniciosi* and the protozoan *Aphytis diaspidis.*
- » *Chilocorus infernalis, Chilocorus rubidus,* and *Pharoscymnus flexibilis* are all predators.

Blister mite, *Aceria fici*

Damage symptoms

- » Leaves get blistered and russetted, branches may become stunted, and leaves may fall from trees.
- » Feeding near the aperture of the eye causes rusty brown spots on the florets
- » The fruit gets damaged upon opening.
- » The affected fruits are not as tasty.
- » Move the fig mosaic virus around.

Pest identification

- » A tiny mite (about 0.003 to 0.005 of an inch long) with two sets of legs near the front of its wedge-shaped, pale yellow body.

Management

- » In the event that you find any damage on the early fruits, throw them away
- » Sulfur or horticultural oil sprays work well as an application.

DISEASES

Rust, *Cerotelium fici*

Symptoms

- » Small, sharp, yellow-green specks first appear on the leaf and are the first sign of the disease.
- » There is no dramatic enlargement of the dots, but they do get more yellow and then a yellowish brown.

- » A crimson rim surrounds the place.
- » The bumps are smooth on the top and look like little blisters on the bottom.
- » At maturity, brown spores are expelled from the blisters.
- » Yellowing and eventual leaf margin death characterise the progression of infection in infected leaves.
- » Death and defoliation are inevitable outcomes.
- » Within a matter of weeks, the trees might be completely bare.
- » In most cases, rust on figs becomes an issue when the fruit is fully ripe.
- » These fungi can also infect fruit.

Favorable conditions

- » The pathogen thrives in warm, humid climates where it frequently rains.
- » The susceptibility of hosts to infection by the algae increases in conditions characterised by low plant nutrition, poor soil drainage, and stagnant air.

Survival and spread

- » Spots on leaves and stems, as well as dead and decaying plant matter, are where the infections multiply and live.
- » Urediospores disperse the pathogen through the air.

Management

- » The dropped and contaminated leaves must be gathered and disposed of.
- » To improve air flow within the foliage, the plant can be trimmed.
- » Avoid watering the leaves directly, as this encourages disease.
- » Mancozeb 0.25%, Zineb 0.20%, or Propiconazole 0.01% spray is suggested. When the first leaves of spring have fully expanded, that's when you should begin applying a fungicide.
- » Two or three times a month, spray a 5-5-50 Bordeaux mixture.
- » The use of neem oil can be effective against a rust infection if caught early.

Leaf spot, *Cercospora fici*

Symptoms

- » Tiny brown spots emerge on the leaves at first, and then they spread and become larger, reddish-brown lesions with dark brown edges, which can be either uniform or zonate.

» Affected leaves fall off sooner and develop uneven patches due to these lesions.

» The veins in the leaf's centre dry out and fall out.

» Mycelium that looks like cobwebs covers the undersides of leaves and eventually turns powdery.

» These fungi can also infect fruit.

Favorable conditions

» Conditions of 25–27° C and 12 hr. of moist leaves promote the disease.

» When it rains, the sickness spreads like wildfire.

» When plants are in their milk and wax stages, infection is more likely to occur if it is hot and rainy.

Survival and spread

» The mycelium of the fungus overwinters on plant debris that was infected and can live for up to 20 weeks in dry soil.

» With a transmission rate of 22.9%, the fungus is carried through plant material

Management

» It is suggested that a spray solution of 0.25 % Mancozeb and 1 % copper oxychloride be used.

Anthracnose, *Glomerella cingulata*

Symptoms

» The fruit and the leaves are equally susceptible to the fungus that causes anthracnose.

» Infected fruit exhibits soft decay and falls from the tree before its time.

» Dried-out, unripe fruit may still be on the tree.

» Small, sunken, discoloured region caused by infection.

» Pink spore mats cover the expanded areas as they age.

» A dark brown ring will form around the edge of affected leaves. Infection rates rise, leading to defoliation.

Favorable conditions

» Rain never stops.

» Approximately between 28 and 30 °C.

» Humidity levels that are very high

Survival and spread

» The conidia on contaminated plant detritus and in the air and the spores themselves were the vectors for the disease's rapid spread.

Management

» When growing figs, cleanliness is of paramount importance.

» Infected fruit and leaves should be thrown away immediately.

Mosaic, *Fig Mosaic Virus*

Symptoms

» Leaf displays a mosaic pattern of light and dark green regions,

» Leaves may have pronounced yellow spots, ring spots, oak leaf patterns, distortion, bronzing, or a lowering of the diffuse mottle and vein cleaning.

» The leaves could be abnormally tiny and misshapen.

» Unfortunately, premature defoliation and fruit drop are rather common.

» Abnormally uneven ripening or bruising of fruit.

Spread

» *Aceria ficus*, a kind of eriophyid mite, is responsible for spreading fig mosaic virus. A single mite's feeding can spread the virus to a healthy seedling.

» Vegetative cuttings are a vector for the virus.

» Another mode of transmission for the virus is through grafting.

Management

» Keep your stock plants virus-free.

» Only multiplying from cuttings produced from stock plants free of viruses is recommended.

» Rootstocks may be selected seedlings.

» The eriophyid mite *(Aceria ficus)* is a disease vector that needs to be eliminated.

» If a fig tree develops symptoms, it must be destroyed immediately.

References

Gowen, S. R. (1995). Pests. In *Bananas and plantains* (pp. 382-402). Dordrech Springer Netherlands.

Stover, E., Aradhya, M., Ferguson, L., & Crisosto, C. H. (2007). The fig: overvie of an ancient fruit. *HortScience*, *42*(5), 1083-1087.

Butani, D. K. (1975). Insect pests of fruit crops and their control-16: fig. *Pesticides*, *9*(11 32-36.

10

Grapevine

INSECT PESTS

Flea beetles, *Scelodonta strigicollis*

Damage symptoms

» Adults and larvae damage leaves by feeding on both their upper and bottom surfaces.

» Adults feed on the inflated buds that grow after cutting vines, transforming them into a "dry sprout" while the grubs feed on roots and pupate in the soil.

» The beetle later feeds on fully developed leaves, leaving behind the telltale slits in the lamina.

» Make little holes in the emerging grape buds with your teeth.

» They spend the daytime tucked away under a layer of flaky bark or a clod of earth.

» Adults make tiny holes in young leaves.

» Grubs wreak havoc on roots.

Pest identification

» **Adult** - Reddish-brown, glossy, and with six spots on the elytra in the adult stage.

» **Grub** – Brownish grub with a blackish top.

Management

» Once you've finished trimming the vines, rake the soil in the basin to let the grubs and pupae out into the sunlight.

» Adult beetles can be shaken loose from vines. Put them in kerosene-wate trays for disposal.

» Spraying Carbaryl 50 WP at 4 g/L, Quinalphos 25 EC at 2 ml/L, c Chlorpyrifos 20 EC at 2.5 ml/L seven days after pruning (at early bud sprou is recommended. After 15 days, reapply any of the aforementioned spray

» Ninety % of beetles died just four days after being sprayed with 0.05⁹ Monocrotophos or 0.15% Carbaryl, and these concentrations remaine poisonous for another 33 and 32 days, respectively.

» Flea beetle was effectively combated by applying 5–7.5 t/ha of neem c castor cakes or 2.5 t/ha of tobacco trash.

Thrips, *Rhipiphorothrips cruentatus, Scirtothrips dorsalis*

Damage symptoms

» After pruning, when young, fragile leaves are ideal for oviposition, thes beneficial insects emerge to begin their crucial feeding and reproductio cycles.

» Nymphs and adults both eat by sucking the leaking cell sap from th underside of soft leaves, flower stalks, and mustard-sized berries.

» Their damage, known as "thrips scab" on mature berries, reduces thei commercial and export worth.

» Therefore, thrips control should always occur prior to blossoming.

» White scorch marks appear on the leaves.

» The absence of fruit clusters on a vine.

» The dropping of the fruit before its time.

Pest identification

» **Nymphs** - drab brown tint with a bright scarlet belly.

» **Adult** - Blackish-brownish lower body and wing coverts in adults

Management

» Assemble rotten produce and dispose of it.

» Take away the trimmings and burn them away from the main field.

» Repeatedly rake the ground.

» Acephate 75 SP at 1 g/L should be sprayed prior to blooming, and then either Methyl demeton 25 EC 0.05%, Dimethoate 30 EC 0.06%, or Carbaryl 50 WP at 4 g/L should be applied after berry formation.

» Thrips populations can be reduced in cold and humid climates by spraying with fungal pathogens like *Verticillium lecanii* or *Beauveria bassiana* at 5 mL or 5 g/L, respectively. This is especially effective when temperatures are between 20 and 25 °C and relative humidity is above 80%.

» Luring thrips into capture with yellow sticky traps.

Mealy bug, *Maconellicoccus hirsutus*

Damage symptoms

» Colonies of nymphs and adults can be spotted on young shoots, where they are sucking the cell sap shortly after hatching.

» Difficulty in eradicating mealy bugs is exacerbated by their long generation lengths and the presence of conspecific species.

» Adults and juveniles both feed on the sap of leaves, stems, and fruits.

» When they do, sooty mould quickly spreads throughout their leaves, stems, and branches from the honey dew they emit.

» The leaves are wrinkling and turning yellow.

» Fruit spoilage.

Pest identification

» **Nymph** - Pinkish and nymphlike.

» **Adult** – At this stage of life, they are a mature pink and waxy white.

Management

» After you've done your initial trimming of the main stem and principal branches, you'll want to remove any flaking bark.

» IIHR swab combination (Carbaryl 6 g + Copper oxychloride 10 g + Neem oil 1 ml + Kerosene 1 ml + sticker 1 ml per litre of water) should be used to dab the trunk.

» Six months later, around the time of pruning, repeat the above procedure.

» Two weeks after pruning, spray with Phenthoate 50 EC at 1 ml/L; repeat the treatment 15 days later.

» Apply Dichlorvos 76 EC at 1 ml/L when the fruit is forming.

» Dust the soil with 20 kg per hectare of Quinalphos to get rid of the ants.

» At a rate of 2 ml/L, spray either Methyl demeton 25 EC or Monocrotophos 36 WSC.

» Use 25 gm of fish oil rosin soap per litre of dichlorvos 76 WSC and spray the area.

» Release 10 *Cryptolaemus montrouzieri* coccinellid beetles per vine.

» Avoid a pesticide residue problem in your harvest by stopping spraying three weeks before picking.

» *Anagyrus dactylopii*, an encyrtid, was responsible for up to 70% of parasitism in March and April despite consistent application of insecticide.

Shot-hole borer, *Xylosandrus crassiusculus*

Damage symptoms

» The grape shot-hole borer is a major pest that tunnels up the main stem of the plant from the ground up.

» It's a type of bug sometimes referred to as a "ambrosia beetle," after the fruit on which they feed. These beetles raise grub-feeding fungus called "Ambrosia" in the shot-hole nests they construct.

» The vine is harmed by the beetle because it is being used as a nesting material on the main trunk.

» The primary stem has numerous pinholes, and a powdery substance is leaking out of them

» Subsequently, the trunk develops sticky exudates, and the plants wilt and turn yellow. In about 15 to 18 months, the vine will die if the insect is not controlled.

Pest identification

» In a dorsal view, the pronotum entirely covers the head, the antennal club seems like it was sliced at an angle, and the whole body is shiny and smooth.

» The crassiusculus of this species measures 2.1–2.9 mm in length and has a robust body; its mature colour is a deep reddish brown that grows darker toward the elytral declivity.

» The males are significantly smaller (1.5 mm in length) and formed differently (a drastically decreased thorax and a usually "hunch-backed") than the females.

Management

» After you've done your initial trimming of the main stem and principal branches, you'll want to remove any flaking bark.

» Apply the IIHR swab solution to the trunk (as mentioned under mealy bug).

» Six months later, around the time of pruning, repeat the above procedure.

» Gently scrape away any gummy exudates with a knife.

» Spray main trunk with dichlorvos 100 EC at a rate of 2.5 ml/l (not on the leaves and vines).

» If there are only a small number of holes, you can inject dichlorvos 100 EC at a rate of 2.5 ml/l and then band the trunks with plastic, polythene, or synthetic gunny sacks for three to five days.

» After taking off the bands, use a cotton swab soaked in the IIHR solution.

» If you're serious, try again in a month.

» Get rid of that vine and start over with some new seedlings if you have to.

» The most successful and economical method for treating the pest was a swab containing dichlorvos at 0.228%, acephate at 0.225%, and carbendazim at 0.30%. X. crassiusculus while protecting non-target creatures as much as possible.

Stem girdler, *Sthenias grisator*

Damage symptoms

» As a pre-ovipositional activity, adult beetles girdle (ringing) around the main stem and branches 15 cm above the ground level at night, killing the plant beyond the cut.

» The girdled part is where the mature beetle lays her eggs. Following hatching, the grubs bore into the dry wood.

» Large amounts of plant damage are caused by girdling.

» The vine's wilting starts at the branches.

Pest identification

> » **Grub** - Grub has a set of large, powerful jaws and a dark brown, bulbous hea
> » **Adult –** Beetle with a white patch in the middle of each elytra; grey
>   color as an adult.

Management

> » To avoid egg laying, remove any loose bark at the time of pruning
> » Remove diseased branches by cutting them off at their base and settin
>   fire to them.
> » Killing the bugs by hand may aid the longhorn beetle's migration.
> » Spray 2 gm of carbaryl per litre of trunk space.
> » The application of a pesticide, such as Chlorpyrifos, using a piece of clo
>   wrapped around the stem.
> » Apply Phosalone 35 EC 0.07%, Quinalphos 25 EC 0.05%, or Carbaryl 5
>   WP 0.1% as a spray for the first time just after pruning, and do so again
>   a couple of subsequent applications.

Berry plume moth, *Oxyptilus regulus*

Damage symptoms

> » Flower buds are webbed by larvae in their first instar.
> » Mature larvae swarm on fruit bunch.
> » The green berry fruit was eaten by a caterpillar.
> » Primarily sustained by its own internal resources.
> » "Stung berries" refer to those that have been contaminated.

Pest identification

> » **Larvae** - tiny, bluish-green or pinkish, and have a red line running throug
>   the middle.
> » **Adult** - Tiny Moth

Management

> » Scoop up the infected leaves and bury them far underground.
> » Pupae can be killed by ploughing in the summer.

» Pheromone traps can be used to capture and kill mature males.

» Use of Bubrofezin and Acetamprid applied locally should be sufficient for controlling this pest.

Bat, *Cynopterus sphinx*

Crop losses

» Bats of three different species cause significant damage to grapevine cvs, ranging from 30-36%. In no particular order: Arka Hans, Arka Shyam, Arka Kanchan, and Bangalore Blue.

Damage symptoms

» The telltale signs of bat damage are vines with half-naked bunches hanging from them and discarded fruit and leaves in the vineyard and neighbouring trees.

» When fruits were fully ripe, bats were most active, and vines bordering open areas suffered the most damage.

» In terms of food, grape berries are a favourite.

» It removes individual berries from a cluster, consumes their juice, and then spits out the fruit's pulp and skin.

» Bats are nocturnal, therefore they tend to fly around the vineyards at night, wreaking havoc on the grape clusters.

Pest identification

» The dorsal fur is a tanner shade of brown.

» The underbelly's fur is much darker than the topside's.

» Young people weigh less than grownups.

Management

» Bat damage was effectively managed by erecting nylon netting around the bower and stuffing briar and twigs into the canopy gap on the bower.

» The damage caused by bats is reduced by having lights set up in the vineyard's interior and periphery and keeping it brightly lit at night.

DISEASES

Downy mildew, *Plasmopara viticola*

Symptoms

» Small, transparent (oil spots), pale yellow spots with irregular boundarie first appear on the upper leaf surfaces of infected plants.

» It appears as a downy fungus growth on the underside of the leaves, jus below the spots.

» Defoliation of severely afflicted vines is easy, preventing fruit ripening an cane maturity and exposing fruits to sunburn.

» The virus causes flower and fruit drop when it infects young shoots, tendril inflorescences, and berries in ideal conditions (20-25°C temperature an 80% relative humidity).

» When there is persistent rain in the first week of December, crops in souther India typically fail completely.

Favorable conditions

» 20–22 °C is ideal.

» Humidity between 80% and 100% relative.

Survival and spread

» Oospores in infected plant parts such as leaves, stems, and fruits allow it t persist. Mycelium can also exist dormantly in diseased twigs.

» Sporangial dispersal by atmospheric vectors.

Management

» A collection of dead leaves, twigs, blossoms, and berries should be burne since they may contain dormant oospores.

» With the right amount of spacing and pruning, you can keep your cane off the ground and promote healthy airflow.

» Spraying a 1% Bordeaux combination of Groundnut oil, Coconut oil, an Neem oil on freshly pruned plants helps keep infections at bay.

» Difolatan, Chlorothalanil, Metalaxyl, and Dimethomorph, all sprayed a 0.2%, once weekly, are all successful when used to treat flushes.

» *Fusarium proliferatum*, a biocontrol agent, is successful against downy mildew in the field.

» *Fusarium proliferatum G6*, a biocontrol agent, is effective against downy mildew when applied after infection (Falk et al., 1996).

» Reportedly resistant cultivars include Amber Queen, Cardinal Champa, Champion, Excelsior, and Red Sultana.

Powdery mildew, *Uncinula necator*

Symptoms

» Leaves, canes, tendrils, blooms, and new clusters of fruit develop powdery spots.

» Powdery spots on leaves grow, and the upper leaf surface becomes dusty.

» In dry weather, the leaves of infected plants may curl upward.

» Affected leaves deform and turn abnormal colours.

» In time, the powdery growth will become a dark grey.

» Stem oxidation to a dark brown tint.

» Infected flowers will lose their petals and produce few fruits.

» A diseased vine will appear pale and stunted.

» Aged berries develop a powdery growth that is easily noticeable.

» The skin of infected berries hardens and fractures, and the fruit itself does not mature.

» Berries that are infected at an early stage usually die off.

Favorable conditions

» Warm, muggy weather with a lot of clouds is ideal.

Survival and spread

» Maintain yourself through the infected plant's latent mycelium and conidia.

» The spores, called conidia, floated through the air and helped it spread.

Management

» At the first sign of disease, spray a mixture of 0.2% wettable sulphur and 0.1% carbendazim or 0.05% folicur over the affected area, followed by a second and third spray 10 to 12 days later.

- » Grapevines were protected from powdery mildew all through the wet season thanks to a bioagent called *Ampelomyces quisqualis* (isolate A-10), which was cultured on nutrient-treated cotton thread and hung from the vines (pycnidiospores of mycoparasite suppressed powdery mildew)
- » Contrasted with grapevine. Extreme resistance is present in Red Sultana, St. George, and 1613.

Anthracnose, *Elsinoe ampelina*

Symptoms

- » Infected berries will initially develop a pattern of dark red dots.
- » These spots later become circular, sunken, and ashy-grey; at their most advanced stage, they are bordered by a dark margin that gives them the appearance of "bird's-eye rot."
- » Spot sizes range from about a quarter of an inch in diameter to about half of the fruit.
- » As well as stems and leaves, the fungus can infect petioles, stomates, veins, and tendrils
- » It's not uncommon for the tender sprouts to be covered in numerous areas
- » It is possible for these areas to join together and girdle the stem, killing off the tips.
- » Petioles and leaves that have been affected by the spots curl and become deformed.

Favorable conditions

- » Climate conditions are warm and rainy.
- » Soils that are low and poorly drained.

Survival and spread

- » Mycelium can remain dormant in the cankers of diseased stems.
- » This ailment can be passed from vine to vine, from cutting to cutting, and even through the air via conidia.

Management

- » The most important step following trimming is to remove and destroy any infected leaves and branches.
- » It is possible to control the disease by spraying the canes and leaves that have been clipped with a solution of 2.5 kilogm of ferrous sulphate and 0.5

pint of sulfuric acid per 4-5 litres of water.

» Spraying bud swell, fruitlet stages, and 21 days after second spray: 1% Bordeaux mixture or 0.25% copper oxychloride or 0.1% Carbendazim or 0.2% Mancozeb or 0.2% Difolatan or 0.2% Chlorothalanil or 0.1% Bitertenol (Baycor) or 0.1% Difenconazole.

» The following varieties of cotton are resistant to muscat leaf blight: Cvs. Angur Kalan, Bharat Early, Delight, Camphor Ash, Naibel, Concord, White Beauty Seedless, Muscat Hamburg (Gulabi), Himrod, Hussaini, Karachi Niagra, Khalili, Niabel, Schuyler White, St. Valior, White Muscat, Large White, Golden Queen, Bangalore Blue, Isabella, and Golden Muscat.

Black rot, *Guignardia bidwelli*

Symptoms

» The leaves develop crimson spots in the shape of a circle, and then the edges turn black.

» Infected pathogens can also spread to branch tips.

» The first signs of the disease on fruit are little, soft, brown patches that quickly spread and stain the entire berry.

» After a week, the rotting berries have shrunk and turned into black, brittle mummies.

» There is a ring of tiny black dots around the edge, depicting the fungus' fruiting bodies.

Favorable conditions

» Unlike in colder and drier climates, black rot may do a lot of damage in warm and humid environments.

» Climates with high humidity and regular rainfall promote the spread of the disease.

» Predicting the spread of disease takes into account a number of parameters, including how wet and warm the leaves are.

Survival and spread

» The infected organism first infects the fruit's stalk or a young leaf.

» Rapid production of pycnidia and dissemination of pycnidiospores by meteoric water.

» They make it through the winter and begin growing again in the spring

» Dripping water from infected areas of the plant can quickly spread the bla rot disease to healthy areas.

Management

» Pick a location where the vines will get plenty of sun and fresh air.

» Keep the vines off the ground with trellising, which decreases the amou of infection by reducing the amount of time the vines spend wet from de and rain.

» It is possible to minimise the inoculum by collecting and destroying infect leaves and berries.

» Select only a few vigorous canes each year during the dormant season a prune the rest.

» To lessen the severity of diseases, spray the plants with a 1% Bordea mixture, 0.25% Ferbam, 0.25% Captan, or 0.25% Chlorothalanil wh the new shoots are 15 to 25 cm long and repeat the spraying before a after bloom.

» Mars, Suwannee, Conquistador, Venus, Saturn, Cabernet Sauvignon, Bl Lake, and Sunbelt are just few of the resistant varieties.

Grey mold rot, *Botrytis cinereria*

Symptoms

» The infection first manifests itself in the subcutaneous tissue of the fru which separates the skin from the flesh. When this happens, the affect area turns a pale brown colour.

» When the fungal infection eats its way further into the flesh, it leaves behir a mushy, liquid mass of rotting tissue.

» The powdery grey mould stage typically appears on the fruit's surface as t fungus sporulates in a humid environment.

» A disease causes infected fruit to shrivel and turn brown. Fruits within infected bunch that is particularly dense may split open as they matur Such infected fruit rots into a cluster.

» It just takes one infected berry from the field to spread the disease durir shipping or storage, causing a catastrophic loss.

Spread

» The fungus can be passed on by touching it.

Management

» Get rid of any leaves that are touching a cluster.

» Do not apply too much nitrogen.

» Proper pruning, weed and sucker management, and placing or eliminating shoots to promote uniform leaf development will all help with air circulation and light penetration.

» Pruning and thinning can reduce humidity surrounding the clusters, which can help prevent grey rot, and thinning berries in dense clusters can lessen the impact of the disease.

» With three applications of 0.2% Captan one month apart just before precipitation, pre-harvest infections can be kept at bay.

» Transportable fruits should be chilled to 4.44 °C, while those destined for storage should be chilled to -0.55 °C to 0 °C.

» The use of packing material containing sodium bisulphate, which decomposes into sulphur dioxide gas (SO_2) when exposed to moisture, has been shown to be effective in preventing the spread of this disease.

» Researchers found that the fungus *Trichoderma harzianum* was highly successful in combating grey rot.

Rust, *Phakopsora euvitis*

Symptoms

» On the upper surface of grapevine leaves, little dark, angular, necrotic lesions can be seen.

» *P. euvitis* causes variable sized and shaped chlorotic and necrotic lesions on the upper leaf surface during its uredinial and telial phases.

» Lesions are accompanied with sporulating pustules on the lower leaf surface.

» There are telltale signs in places where yellow or orange spores are packed tightly inside pustules on the underside of the leaves.

» Bring on early leaf drop and a drop in fruit quality and production.

Spread

» Rusts reproduce by releasing urediniospores and basidiospores into the air respectively.

» Desiccation, through the thin spore walls, may be able to stop the long distance spread of urediniospores.

Management

» In the months of July and August, as well as January and February, apply three to four sprays of Baycor at 0.1% or Chlorothalanil at 0.2% every two weeks.

Cane and leaf spot, *Phomopsis viticola*

Symptoms

» Lesions on the vine's stem and leaves may be an indicator of a problem.

» Grapes frequently suffer disease loss due to rachis and berry infections.

Management

» In order to get to the good tissue, dead canes or limbs need to be trimmed back far.

» Bordeaux paste should be used to seal cut ends.

» The canes that have been cut back should be gathered and burned.

» The disease rate can be lowered by spraying a 1% Bordeaux mixture every two weeks.

» Depending on the severity of the infection, Mancozeb, Captan, and Ziram all do a great job of keeping it from spreading to the rest of the plant's vascular system.

Pierce's disease, *Xylella fastidiosa*

Symptoms

» Eventually, the leaf margins dry up or die in concentric zones, turning dull yellow or red depending on the variety.

» The petiole (leaf stem) remains attached to the cane after the leaf has dried and fallen.

» The development of the wood in young canes is erratic, resulting in splotches of green surrounded by darker, more fully developed bark.

» Raisins are dried fruit clusters.

Spread

» Disease is spread by snipers.

Management

» Do not let pests like mosquitoes and rodents into vineyards from neighbouring areas, especially in the spring.

» It is also important to remove infected vines from a vineyard as soon as possible once the first signs of Pierce›s disease develop.

» The number of vector insects that migrate into vineyards every spring has been drastically reduced thanks to insecticide treatments in the areas surrounding the vineyards, hence reducing the prevalence of Pierce's disease.

Bacterial canker, *Xanthomonas campestris* pv. *viticola*

Symptoms

» It manifests as tiny water-soaked spots on the undersides of leaves, encircled by a yellowish halo.

» Eventually, these blemishes will grow in size and take on a more angular, dark brown colour.

» At times, a number of smaller patches will merge to produce a larger one.

» Infected veins in the leaves are also visible.

» On petioles and canes, lesions are brown to black, elongated, and cankerous.

» Canes exhibit symptoms like stunting, cracking, and irregular growth at an early stage of disease.

» Brown to black, cankerous lesions develop on berries, and the seriously damaged berries shrink and dry up.

» In most cases, cankerous symptoms appear on every portion of the plant, hence bacterial canker was chosen as the new name for this disease.

Favorable conditions

» It seems that the ideal temperature for the progression of the sickness is between 25 and 30 °C.

» The presence of free water on the leaf, whether from dew, irrigation, or rain, is more crucial to the onset of disease.

» Extremely wet periods saw an uptick in the spread of disease.

Survival

» The bacterium can hang around in infected dry leaves for up to 65 days and in wet soil for up to 35 days.

» Alternate hosts could come from Mango, Neem, or *Phyllanthus maderaspatensis*

Management

» Planting disease-free cuttings helps lower disease rates.

» Killing out contaminated plant tissue lessens the overall population of the disease

» It is important to limit the amount of water used for irrigation.

» Pruning in late October can help prevent canker.

» Antibiotics sprayed at 3000 ppm beginning in the second leaf stage and repeated every 15 days effectively suppresses the disease.

Crown gall, *Agrobacterium vitis*

Symptoms

» The most prominent symptom of this disease is the development of galls on the vine's aerial sections.

» Galls manifest as bumps on the trunk and roots of a vine.

» Grape galls on roots are uncommon, however bacteria can cause necrosis of the roots in specific areas.

» To the touch, young galls range in colour from creamy white to a sage green and they have no protective bark or covering.

» As time passes, the tissue turns a brownish colour.

» As the surface opens up, it gains a somewhat hard and highly rough texture

» Although the galls' dead outer tissue gives the impression that the bacterium is also dead, it is actually still active within the vine.

» Development of galls can cause girdling of vines, which can lead to decreased vigour and productivity.

Management

» The number of Agrobacterium and nematodes in soil is decreased by solarization.

» It's important to exclusively use in vitro apical meristem or shoot-tip tissue culture-derived planting substrate verified free of known pathogens.

» For grafting, dormant cuttings can be treated with heat (hot water submersion) or chemicals (Oxyquinoline sulphate).

» Take out and throw away any diseased plants or other items.

» Soil burial and mulching can shield a graft union from potential harm.

» High-frequency, multi-trunk training systems for constant trunk replacement

» Roots of cuttings are submerged in a bacterial suspension (*A. radiobacter*) before being transplanted into the fields, which helps to control *A. vitis* with its cousin *Agrobacterium radiobacter* strain K84 (Commercial formulations: Galltroll-A, Nogall, Diegall, and Norbac 84C).

Fanleaf disease, *Grape Fanleaf Virus* (GFLV)

Symptoms

» Chlorotic mottling, such as a yellow mosaic or veins, is another sign that may appear on leaves.

» The irregularly shaped leaves have large petiole sinuses and an aberrant vein pattern, giving the plant an open fan shape.

» Extreme distortion, asymmetry, cupping, puckering, and sharp dentations will characterise the leaves.

» Yellow mosaic, another symptom, can spread to all regions of a vine and be fatal.

» Fruit from GFLV-infected plants will be softer and smaller than that of healthy plants, and the plants themselves will be smaller.

» Reduced fruit set and shrivelled berries characterise these little bunches.

» Unpredictable ripening of bunches.

Spread

» Dagger nematodes, *Xiphinema index* and *X. italiae*, are responsible for their transmission.

» The virus and its vector can travel great distances thanks to infected spreading material.

Management

» The use of verified disease-free plant material.

» Wheat and lucerne should be rotated every seven years.

» To combat the nematodes that spread this disease so easily, it is recommended to replant grape beds heavily with nematicidal plants like French marigolds

» Field resistance is provided by the grapevine rootstocks 039-16 (*V. vinifera* × *V. rotundifolia*) and 043-43.

References

Bettiga, L. J. (Ed.). (2013). Grape pest management (Vol. 3343). UCANR Publications

Flaherty, D. L., Jensen, F. L., Kasimatis, A. N., Kido, H. and Moller, W. J. (1981). Grape pest management. Agricultural Sciences Publications, University of California.

Halleen, F. and Fourie, P. H. (2016). An integrated strategy for the proactive management of grapevine trunk disease pathogen infections in grapevine nurseries. South African Journal of Enology and Viticulture, 37(2), 104-114.

Koledenkova, K., Esmaeel, Q., Jacquard, C., Nowak, J., Clément, C. and Ait Barka, E. (2022). Plasmopara viticola the Causal Agent of Downy Mildew of Grapevine From Its Taxonomy to Disease Management. Frontiers in Microbiology, 13, 889472.

Mondani, L., Palumbo, R., Tsitsigiannis, D., Perdikis, D., Mazzoni, E. and Battilani, P. (2020). Pest management and ochratoxin A contamination in grapes: A review. Toxins, 12(5), 303.

Wan, Y., Schwaninger, H., He, P. and Wang, Y. (2007). Comparison of resistance to powdery mildew and downy mildew in Chinese wild grapes. VITIS GEILWEILERHOF-, 46(3), 132.

Wilson, H. and Daane, K. M. (2017). Review of ecologically-based pest management in California vineyards. Insects, 8(4), 108.

11

Guava

INSECT PESTS

Green shield scale, *Pulvinaria psidii*

Damage symptoms

- » Feeding on the leaves or tender twigs and shoots weakens the host and leaves sticky honeydew deposits on nearby surfaces. These small, sessile, oval-shaped scales are greenish-yellow in colour.

- » Aphids, both nymphs and adults, feed on leaf sap.

- » Leaves are turning yellow.

- » Some ants might be drawn in by the honeydew.

- » The sugar deposits quickly become covered in a black mould.

- » The quality of fruits can suffer and the leaves may fall off too soon if they are badly soiled.

- » Each adult scale on a stem will produce a white wax mealy coating and a white wax ovisac that is as long as or longer than the insect.

- » The summer is when the infestation is at its worst.

Pest identification

- » **Nymph:** Yellow as a nymph's pale cheeks.

- » **Adult** – In its adult form, this soft scale fish is a vibrant shade of green.

- » The 3 mm long adult is oval in shape, yellowish green, and shaped like a shield.

Management

- » Remove damaged branches and burn them during the dormant period.
- » If you catch an infestation early enough, you can just cut down the infected shoots and throw them away.
- » To stop the spread of pests to uninfected trees, remove all limbs that touch the afflicted one.
- » The sooty mould can be sprayed away using 2% starch spray.
- » For severe infestations, cut off infected pieces and spray with either Diazinon 40 EC at 2 ml/L, Quinalphos 25 EC at 2 ml/L, or Monocrotophos 36 SI at 1.5 ml/L.
- » Spread 25 gm of 2% methyl parathion dust around the tree's base to kill any ants that might try to climb it.
- » Good results in controlling P. psidii were achieved by using a combination of acephate and demeton-S-methyl.
- » January/February is the best time to release 20 *Cryptolaemus montrouzieri* grubs per tree.
- » Foster the development of parasitoids such as *Aneristis sp., Coccophgagus cowperi*, and *C. bogoriensis*.
- » The most important natural opponent of *P. psidii* is the *microteryx kotinsky* and it has been claimed that it has completely biologically controlled the pathogen.

Bark eating caterpillars, *Indarbela quadrinotata, I. tetraonis*

Damage symptoms

- » Nests are made in the bark's fissures and divots.
- » Larvae, as they hatch, eat holes in the bark or the main stems.
- » Caterpillars spend the daytime underground and emerge at night to dine on bark.
- » Caterpillars are able to bore through quite thick tree bark, including the trunk, the point where the main stems join, and the thick branches.
- » They live in the silken tunnels filled of frass and faecal materials on the bark's surface, where they may eat and hide from predators.
- » Galleries are used for the pupation process.

» Newly planted saplings may not be able to withstand the assault.

Pest identification

» The adult stage of this moth is characterised by thick, yellowish-brown wings that are marked with brown, undulating lines.

» Men are noticeably shorter than women.

Management

» Protect your orchard from borers by keeping it clean.

» Cut off and dispose of the tree's dead or seriously damaged limbs.

» Eradicate the use of any substitute hosts, including silk cotton.

» An iron spike can be inserted into the holes to kill the caterpillars mechanically if an infestation is discovered.

» Rub a mixture of coal tar and kerosene (1:2) or carbaryl 50 WP (20 g/L of water) over the lower three feet of a tree's trunk.

» Adult beetles cannot lay their eggs in the loose bark, so it is scraped off.

» Wet a cotton swab with Dichlorvos 100 EC @ 2 ml/L and insert it into the holes after cleaning the damaged area of the trunk or main stem.

» Cotton soaked with Monocrotophos 36 WSC at a rate of 10 ml per 2.5 cm each tree serves as padding.

» Using a hook, remove the grub from the bore hole, and then fill it with 10 to 20 ml of Monocrotophos 36 WSC.

» Fill holes with mud and 5 gm of Carbofuran 3G.

» Using a 0.1% Aldrin spray, cover the tree and its branches.

Mealy bugs, *Ferrisia virgata, Planococcus citri, P. lilacinus, Maconellicoccus hirsutus*

Damage symptoms

» Many guava orchards in South India suffer extensive damage from mealy bugs, specifically *Ferrisia virgata, Planococcus citri, P. lilacinus,* and *Maconellicoccus hirsutus.*

» They cause substantial economic losses and are found on both young plants and ripe fruits.

» Reduced demand and lower prices are the results of fruit contamination by mealy bugs and sooty mould.

» Fruit ripening too quickly.

Pest identification

» **Nymph** - may range from very light yellow to almost white.

» **Adult** - Females reach maturity as elongated, slender beings covered in a white waxy substance.

Management

» Neem oil 5 ml/L and Triozhophos 2 ml/L or Phosalone 35 EC 1.5 ml/L and Neem oil 5 ml/L are both effective sprays.

» Let the parasites out, Predators, including *Aenasius advena* and *Blepyrus insularis,* The scymnids *Scymnus coccivora, Mallada boninensis,* the shrew *Brumus suturalis,* and the eagle ray *Spalgis epius.*

» The encyrtid *A. advena* is a major parasitoid of *F. virgata,* responsible for as much as 50% of field parasitism.

» Parasitism due to *Blepyrus insularis* can reach 32 %.

» In order to rid trees of *P. lilacinus* and *M. hirsutus,* it is recommended to release 10 adult beetles of *Cryptolaemus motrouzieri.*

» The population of P. citri is controlled by parasitoids like *Coccidoxenoides perigrinus* and *Allotropa citri* and predators like *S. epius* and *C. montrouzieri*

» The parasitoid *Leptomastix dactylopii* also provides efficient and long-lasting control of *P. citri.*

Fruit flies, *Bactrocera diversus, B. dorsalis*

Damage symptoms

» In the months between September and May, when guava fruits are in season, the fruit flies feast on them.

» They have a range of 1-2 kilometres and can fly to find a host.

» Semiripe fruits are a favourite snack for adults and larvae.

» Fruits typically show punctures from oviposition.

» Maggots eat pulp and turn it into a foetid, semi-liquid mess.

est identification

» In contrast to the brownish black of *B. dorsalis*, which lacks a yellow middle stripe on its thorax, the adults of *B. diversus* are a smoky brown colour with greenish black thorax with yellow markings.

Ianagement

» Steer clear of acquiring crops during the rainy season, since the occurrence rate is really high.

» Get rid of the fruit that has fallen from the tree but has fruit flies on it.

» Expose and kill pupae by ploughing during the summer.

» For monitoring and eliminating adult fruit flies, set out a Methyl eugenol lure trap (25 per hectare). Mix 10 ml of methyl eugenol with 50 ml of malathion each trap to get a 1: 1 combination.

» Both Phophamidon and Oxydemeton methyl, at concentrations of 0.03%, can be used as an effective spray for exterminating the insect.

» Malathion EC 0.05% Spray.

» Molasses or jaggery at 10 g/L combined with Malathion 50 EC 2 ml/L or Dimethoate 30 EC 1 ml/L should be sprayed twice, 14 days apart, before the fruits reach full ripeness.

» Dissemination of parasitoids in the wild, such as *Opius compensates, Spalangia philippinensis*, and *Diachasmimorpha kraussi*.

piraling whitefly, *Aleurodicus dispersus*

amage symptoms

» Aphids, both nymphs and adults, feed on leaf sap.

» They produce honey dew, which feeds the fungus that causes sooty mould.

» Leaves are turning yellow.

est identification

» **Nymph** - thin, waxy rods that stick out from the body like short glasses.

» **Adult** - dazzling white and most active in the wee hours of the morning.

Ianagement

» Reducing contamination in the field.

» Elimination of potential parasite hosts.

» Yellow sticky traps have been set up.

» Imidacloprid 200SL at 0.01% or Triazophos 40EC at 0.06% should be use during high infestation.

» Spray either NSKE 5% or 3% Neem oil.

» Coccinellid predators, including *Cryptolaemus montrouzieri* , *Axinoscymn puttarudriahi*, and *Mallada astur*, are set free.

» The parasitoids *Encarsia haitiernsis* and *E. guadeloupee* are being released

Fruit borers, *Deudorix (Virachola) isocrates*

Damage symptoms

» Up to 10% of guavas were affected by this pest.

» A caterpillar eats its way through an apple.

» Consumes its own flesh (pulp and seeds)

Pest identification

» **Larvae** - covered in short hairs and are a dark brownish colour.

» **Adult** - butterflies are a rosy brown colour. Females can be identified by th v-shaped patch on their front wings.

Management

» Remove spoiled produce from storage.

» Weed plants are used as a substitute in organic farming.

» Monitor adult behaviour with a light trap set up at a rate of 1 per hectar

» Two applications of Malathion 50 EC 0.1% spray should be made, on during bloom and again just before fruit sets.

» If damage is more than 5%, spraying with 0.2% Carbaryl every 10 to 1 days during June and July can help.

Tea mosquito bug, *Helopeltis antonii*

Damage symptoms

» Petioles, young shoots, and fruits are all punctured by nymphs and adul alike.

» Leaf tissue develops brownish-black necrotic spots.

» On the shoots, long streaks and spots appear.

» Fruits develop corky scabs

» There was evidence that the tea mosquito bug could damage guava cvs. Lucknow-46, Lucknow-47, Bangalore, and Red Fleshed.

Pest identification

» Reddish brown nymphs and adults have a black head, a red thorax, and a black and white abdomen.

Management

» Gather up the broken pieces of the plant and throw them away.

» If you want your flowers to last as long as possible, you should spray Malathion 50 EC 0.2% every other month while they're in bloom.

» Difference between guava and. Smooth Green was less susceptible than Safeda and Seedless.

» *Beauveria bassiana*, a fungal pathogen, was sprayed once weekly at a spore concentration of 1×10^9 spores per ml with appropriate adjuvants, and the resulting reduction in pest damage was comparable to that seen with chemical pesticides.

DISEASES

Wilt complex, *Fusarium oxysporum* f. sp. *psidii*, *Helicotylenchus dihystera*

Symptoms

» The leaves become dark, wither, and fall to the ground.

» Both nitrogen and zinc are lacking in the diseased leaves.

» Normal die-back symptoms can be seen in the infected branches and twigs.

» There is a noticeable discoloration in the central portion of the stem and root, all the way to the cambium and vascular tissues.

» There has been no increase in fruit size.

» The entire plant can be lost in as little as three to four weeks as the disease progresses.

» Veins in the root's xylem have turned a characteristic black in this cross section.

Favorable conditions

» Heavy downpours in the months of August and September.

» A prolonged period of water stagnation in the field.

» We recommend between 23 and 32 °C, with a relative humidity of 76%.

» Soils with a high alkalinity (pH 7.5–9.0) or a low acidity (pH 6.5) are idea

» Younger plants, especially those less than 3 years old, are more at risk.

Survival and spread

» Plants contaminated with disease can spread it to undeveloped regions
they are moved there.

» Messages can travel a short distance via water.

Management

» Using sunlight to stimulate plant growth in soil.

» Correct field hygiene procedures.

» Cut down and burn the wilting vegetation.

» Marigolds and turmeric, when planted in an intercrop, proved effective i
preventing wilt.

» Harsh trimming, then four applications of 0.2% Benlate or Bavistin every yea

» Half of the anthracnose disease that occurs on fruits after harvest can b
prevented with *Bacillus subtilis*.

» *Aspergillus niger* AN 17 (Pusa Mrida), *Penicillium citrinum*, *Trichoderm
harzianum*, and *T. virens* were determined to be the most efficient agent
for biological control.

» Incorporation of drought- and salt-tolerant cvs like Allahabad Safed:
Dholka, and Sindh.

Anthracnose, *Colletotrichum gloeosporioides*

Symptoms

» Loss of Branches

» Sunken, dark-colored, necrotic lesions are typical signs.

» Unripe fruit exhibits pin-sized dots, which grow to be 5-6 mm in diamete
as they ripen.

» Budding flowers and buds are also prey, leading to their premature dropping.

» Damaged areas of unripe fruits become corky and rigid, and in the case of serious infection, they sometimes fracture.

» The necrotic sores become blanketed in pink spore masses when exposed to high humidity.

» Spores form as the small sunken lesions join together to form larger necrotic areas in the fruit's flesh as the disease advances.

Favorable conditions

» Planting closer together without thinning the canopy.

» Having thinning, trimming, or harvesting done and then leaving fruit and leaves on the ground.

» Spore formation and canopy-wide dissemination are both aided by dew and rain.

» Diseases thrive best in temperatures between 10 and 35 °C (with 30 degrees being optimal) and in air that is 95 % humid.

Survival

» The principal source of inoculum is the infected fruits, twigs, and branches.

» By spores carried on the wind from the orchard's decaying leaves, twigs, and mummified fruits.

Management

» The infected limbs, fruits, and leaves must be cut down and burned as soon as possible after the wet season ends.

» As soon as the disease is spotted in the orchard, a spray of 1% Bordeaux mixture, 0.2% Difolatan, 0.2% Zineb, or 0.1% Topsin M should be applied.

» Sodium metabisulphate or bleaching powder can be used to fumigate the fruit after harvest.

» The apple guava (white fleshed variety) has only moderate resistance to anthracnose.

» In terms of pre-inoculation sprays, Thiram, Kirti-Copper, and Cuman performed admirably.

» Tetracycline dip therapy used after inoculation was successful in suppressing the disease.

» Fruit irradiated with 100 Kr gamma rays for 5, 18, and 24 hours.

Fruit rot, *Phytophthora nicotianae* var. *parasitica*

Symptoms

» Any part of a ripe fruit's surface, including the flower and stem end, can be the origin of a disease.

» As the fruit ripens, a white cottony growth rapidly appears and, under humid conditions, can cover practically the whole surface in as little as 3-4 days.

» Young, immature fruits are more susceptible to disease. In time, they turn brittle and woody.

» Most severely impacted are fruits that are close to the soil, are shaded by heavy foliage, and are exposed to high relative humidity.

» The fruit's skin softens a bit behind the cottony white growth, darkens in colour, and develops a distinctive foul odour.

» These fruits either remain whole or drop off.

Favorable conditions

» Protect the plantations.

» Climates that are cool and damp, with high soil moisture levels, are conducive to the growth of several diseases.

» It is easier for diseases to start when conditions are damp and warm (between 28 and 32°C; 25°C is ideal), the soil is poorly drained, and there have been accidents.

Survival and spread

» The inoculum can be maintained in contaminated soil or plant matter.

» Spores can spread more easily in wet and windy conditions.

» Spores in diseased soil or vegetation are carried by raindrops to new locations.

Management

» Using sunlight to stimulate plant growth in soil.

» Elimination of dead vegetation.

» For best results, spray once a week with a 2: 2: 50 Bordeaux combination or 0.2% Copper oxychloride.

» When fruiting, spray once a month with 10 ppm Aureofungin or 0.2% Zineb.

Rust, *Puccinia psidii*

Symptoms

» Disfigurement, defoliation, stunted development, and, in extreme cases, death are all common manifestations of this disease.

» If the infection is severe, it can cause widespread flower blight, fruit infection, defoliation, and twig dieback.

» Guava's leaves, stems, flowers, and fruit can all fall victim to the infection.

» It's possible for infected fruit tissues to deform and be riddled with bright yellow pustules.

» Dark-bordered, roughly round brown lesions with yellow halos appear on fully grown leaves.

Favorable conditions

» Warm temperatures and heavy humidity encourage the spread of disease.

Management

» It is recommended to conduct field scouting for the onset of disease or at seasons of the year when environmental conditions are favourable for pathogen infection, in order to apply fungicides at the appropriate times.

» A healthy, aggressively growing tree is less susceptible to disease stresses, and proper cultural practises like fertiliser, watering, trimming, and sanitation all help to achieve this.

» Cut down diseased branches and throw away infected leaves.

» As soon as rust emerges, use a fungicide such as Copper hydroxide plus Mancozeb, Azoxystrobin, or Chlorothalonil.

» When possible, "correct" irrigation procedures and plant spacing can reduce the length of time that relative humidity is high and leaf wetness persists.

Scab/fruit canker, *Pestalotia psidii, Colletotrichum gloeosporoides*

Symptoms

- » Necrotic patches appear as tiny, intact circles of brown or rust colour; the spots eventually tear open the epidermis in an acircinate pattern durir the later phases.
- » It has a raised periphery and a low centre.
- » Crater-like depressions are more obvious on fruits than they are on leave
- » Fruits split open and seeds are exposed when cankerous, elevated patch proliferate.
- » Fruits that have been infected with the fungus will fail to mature normal develop deformities, and eventually shrivel and die.

Favorable conditions

- » At 25-30°C with high relative humidity, the disease is at its worst.

Management

- » Removing and disposing of infected plant material.
- » Disease can be contained in its early stages by spraying a 1% Bordeau mixture at a rate of 2-3 litres per tree every 15 days.
- » Mancozeb sprays containing 0.2% of the antifungal are used as a preventati measure to control disease.
- » Fruits washed in 200 parts per million of salt after harvest The fruits ca be protected from spoilage by aureofungin for up to five days (Kaush et al., 1972). Aretan fruit dip is another method for preventing fruit rc Treatment with this compound is also effective in preventing fruit rot caust by *Colletotrichum gloeosporoides* (Khanna and Chandra, 1976).
- » For canker resistance, Nasik was nearly immune, Safeda and Apple Col were very resistant, Sind was resistant, and Dholka was just mildly resistar

Bacterial blight, *Erwinia psidii*

Symptoms

- » Leaf drop and twig dieback may be caused by large necrotic lesions with translucent halo at the leaf margins or by little water-soaked areas, occasional with a chlorotic halo.

» The xylem is a major transport route for bacteria that can infect other parts of the tree.

» In extreme circumstances, trees lose their leaves and die.

Management

» The most effective treatments were the use of 5.0 g Copper sulphate (SC), 3.5 g Copper oxychloride (COC), 3.0 g Copper hydroxide (HC), and 20.0 mL liquid bioactive compost (CBL) per litre of water.

Algal leaf spot, *Cephaleuros virescens*

Symptoms

» Orange, rust-colored, dense silky tufts, 5–8 mm in diameter, are seen on both the abaxial and adaxial sides of infected leaves.

» When these spots are removed, a thin, greyish white to dark necrotic crust is left on the leaf.

» On a leaf, these blemishes tend to cluster together, creating huge, erratic areas. The dots grow to a drab grey green tint.

» The pathogen's filaments spread and expand into the cortical tissues of the host, causing the bark to crack on twigs and branches as well.

» The fungus is also responsible for fruit spots.

Spread

» The disease is more likely to spread in wet, humid weather.

» Splashing water about can disperse zoospores.

Management

» Cultural practises such as weed management, fertilising, irrigation, pruning to increase canopy airflow and light penetration, and wider spacing between trees all contribute to keeping trees healthy and vigorous.

» To prevent algae growth, use a Bordeaux mixture or fungicide containing copper on a regular basis.

» The vitality of trees and their resistance to this disease can be improved through the control of insect, mite, and other foliar diseases.

References

Abbas, M., Hussain, D., Saleem, M., Ghaffar, A., Abbas, S., Hussain, N. and Ghaffar A. (2021). Integrated pest management of guava, citrus and mango fruitflies a three districts of Punjab. Pakistan Journal of Zoology, 53(3), 995.

Gould, W. P. and Raga, A. (2002). Pests of guava. In Tropical fruit pests an pollinators: biology, economic importance, natural enemies and control (pp 295-313). Wallingford UK: CABI Publishing.

Gundappa, B. and Mani, M. (2022). Pests and Their Management in Guava. Trend in Horticultural Entomology, 605-623.

Kalita, M. K. and Srivastava, J. N. (2022). Guava (Psidium guajava): key diseases an their management. Diseases of Horticultural Crops: Diagnosis and Managemen Volume 1: Fruit Crops, 217.

Khan, M. M., Shah, S. H., Akhter, I. and Malik, H. (2017). Integrated pest managemer of fruit flies in guava orchids. Journal of Entomology and Zoology Studie 5(2), 135-138.

Lim, T. K. and Manicom, B. Q. (2003). Diseases of guava. Diseases of tropical frui crops, 275-289.

Misra, A. K. (2004). Guava diseases—their symptoms, causes and management. I Diseases of Fruits and Vegetables: Volume II: Diagnosis and Management (pp 81-119). Dordrecht: Springer Netherlands.

Misra, A. K. (2005, December). Present status of important diseases of guava i India with special reference to wilt. In I International Guava Symposium 73 (pp. 507-523).

12

Litchi

INSECT PESTS

Fruit and shoot borer, *Conopomorpha sinensis*

Damage symptoms

» The larvae eat their way into the flesh of the plant.

» Insects cause failure of shoot growth in September and October, especially in freshly established orchards, by damaging the newly developed shoot.

» In April and May, it pierces the peduncle of developing and ripening fruits, causing severe loss due to premature fruit drop and the emergence of excreta/larvae when the fruit is cut/opened.

» They bore right through the juicy pulp of the fruit, which is why so much of it ends up on the ground.

» The most obvious stage for diagnosing borer incidence in the field is the adult stage, when the larvae emerge from the fruit and pupate on the leaf surface under a spinning cocoon.

Management

» Systemic insecticides such as Thiacloprid 21.7 SC or Imidacloprid 17.8 SL @ 0.5-0.7 ml/L may be treated at 15-day intervals throughout the month of September.

» Neem oil (4 ml/L) sprays can be applied prior to flowering to prevent egg laying.

» After fruit has reached the clove size, a spray of either Lufenuron 5 EC (0.7 ml/L) or Novaluron 10 EC (1.5 ml/L) is applied (at 10-12 days after fruit set).

» 25-30 days after fruit set, when the aril (pulp) is developing, spray with Cypermethrin 5 EC @% 0.5 ml/L or Emamectin benzoate 5% SG (0.4 ml/L)

» Ten days prior to harvesting, apply a last spray of Novaluron 10 EC 1.5 ml/L or Cypermethrin @% EC 0.5 ml/L.

» Trichogramma chilonis is an example of a parasitoid.

» Mirid bugs (*Campyloneura sp.*), ladybird beetles (*Cheilomenes sexmaculata*, *Coccinella septempunctata*-seven spotted, *Brumoides suturalis*-three striped), lacewings (*Chrysoperla carnea*), big-eyed bugs (*Geocoris sp.*), and pentatomic bugs are only few of the insects that feed on Mirid bugs (*Eocanthecona furcellata*).

Leaf curl mite, *Aceria litchii*

Damage symptoms

» Mite nymphs and adults alike cause harm by draining the cell sap from fresh leaves, flower buds, and ripening fruit.

» Repeated sucking causes irritation in leaf tissues, which develops into erineum.

» Initially appearing as a velvety growth on the bottom leaf surface, the condition is identified by its gradual expansion into a brown-chocolate color, deep lesion on the entire leaf, and subsequent curling, all of which contribute to a decrease in the leaf's photosynthetic area.

» The erineum begins life as a silvery white, then darkens via brown and reddish brown to black.

» If the deformity is severe enough, the entire terminal may be distorted.

» Unmanaged litchi orchards have been found to have extremely high rates of infestation, which can quickly spread to nearby plants and orchards. Low levels of flowering and fruiting in such an orchard result in substantial financial losses for the farmers.

» In un-pruned and poorly maintained orchards, the mite is most prevalent between the months of July and October, and again in February and March.

» If infestations are bad enough, many leaves will be lost.

» This is harmful to developing fruit trees.

» Moving from the leaves to the growing flowers and fruits can also be problematic for the plant.

» Both fruit set and fruit deformation are possible.

Favorable conditions

» The highest mite population was found in April and May, and the lowest was found in November and December, when temperatures and relative humidity were at their lowest.

Survival

» The adult mites survive the winter.

Management

» Only plants that have not been infected should be used to make the layers.

» After fruit has been harvested but before a new flush emerges, the plant is pruned and infected branches are removed, and two sprays of Chlorfenapyr 10 EC (3 ml/l) or Propargite 57 EC (3 ml/l) are applied at 15-day intervals throughout the month of July.

» Also, in October, cut back any newly afflicted branches and spray them with Chlorfenapyr 10 EC (3 ml/l) or Propargite 57 EC (3 ml/l).

» The aforementioned acaricides should be sprayed as needed after the panicle forms but before the flowers open, in case the mite is present.

» Chrysoperla zastrowii sillemi, anthocorid bugs, predatory mites (Amblyseius fallacis), and coccinellid flies are all potential threats (Stethorus punctum).

DISEASES

Anthracnose, *Colletotrichum gloeosporoides*

Symptoms

» Infected fruit is worthless despite the fact that it causes damage to leaves, branches, flowers, and flower stems.

» Leaf lesions can be either little, circular, light grey patches or larger, irregular brown stains at the leaf tips.

» However, diseases are easily visible on the blooms and fruit.

» Disease caused by the fungus may not always manifest itself until after harvest.

» When the disease initially manifests, it looks like little brown freckles, and it most often emerges on the top of almost ripe fruit in places shaded by overhead branches.

» This time the spots don't grow, but they do become black very quickly.

» At harvest, the spots spread to the fruit's top and sides, eventually coverir up to half of the fruit's surface.

Favorable conditions

» Epidemics tend to occur following periods of warm, wet weather.

Survival and spread

» Infections can survive the winter on the leaves and spread from nurseri to newly planted orchards.

Management

» Use a spray containing either copper oxychloride (0.25%), carbendazi (1%), difenconazole (0.5%), or azoxystrobin (0.023%). Spraying fungicid on crops before harvest can increase their usable shelf life after harvest.

Brown blight, *Peronophytophthora litchi*

Symptoms

» Leaves and panicles susceptible; can infect fruit right up to picking time

» As a result of these infections, output and storage life are shortened.

» Panicles of flowers are very vulnerable.

» Brown spots appear on immature fruit, and white mildew spreads througho the skin of infected fruit before harvest.

Favorable conditions

» Infection is more likely to spread when the weather is consistently dan and warm, between 22 and 25 °C.

Survival and spread

» Fungus spores survive the winter in the soil or on rotting diseased fruit ar are carried by the wind, rain, and insects.

Management

» After harvest, growers should clean up their orchard by cutting down ar shade trees, diseased limbs, or dead wood.

» Copper sulphate in the spring and copper oxychloride in the winter a both effective in reducing inoculum levels.

» During blooming and fruit set, these compounds are swapped out for Fosetyl-AL or Metalaxyl.

References

Cai, Y., Qi, W. and Yi, F. (2023). Smartphone use and willingness to adopt digital pest and disease management: Evidence from litchi growers in rural China. Agribusiness, 39(1), 131-147.

Situ, J., Zheng, L., Xu, D., Gu, C., Xi, P., Deng, Y. and Jiang, Z. (2023). Screening of effective biocontrol agents against postharvest litchi downy blight caused by Peronophythora litchii. Postharvest Biology and Technology, 198, 112249.

Srivastava, J. N., Singh, A. K., Sharma, R. K. and Singh, V. B. (2022). Diseases and Physiological Disorder Spectrum in Litchi (Litchi Chinensis Sonn.)/Longan (Dimocarpus Longan Lour.) and Their Management. Diseases of Horticultural Crops: Diagnosis and Management: Volume 1: Fruit Crops, 271.

Srivastava, K. and Choudhary, J. S. (2022). Pests and Their Management in Litchi. Trends in Horticultural Entomology, 719-734.

Wang, S., Zeng, X., Wang, X., Chang, H., Sun, H. and Liu, Y. (2022). A survey of multiple pesticide residues on litchi: A special fruit. Microchemical Journal, 175, 107175.

13

Mango

INSECT PESTS

Fruit fly, *Bactrocera dorsalis*

Damage symptoms

» The female lays her eggs in clusters within the mesocarp, just below the fruit epidermis (1-4 mm deep), by using her pointed ovipositor to puncture the outer wall of the ripe fruits.

» When these maggots hatch, they feed on the fruit pulp, causing the afflicted fruit to decay and eventually fall to the ground.

» Therefore, the afflicted fruits begin to deteriorate and a brown patch emerges around the oviposition site. Maggots emerge from infected fruit and pupate in the surrounding soil.

» Fluid seepage and dark, rotting spots on fruit.

» The fly population booms in the summer because it may feast on ripe fruit for breeding (May to July).

» After peaking in August, the population gradually begins to fall in September

» Infestations of maggots cause semiripe fruits to rot and fall from the tree before they are fully ripe.

Pest identification

» **Larva** - are apodous maggots that are a dull yellow in colour.

» **Adult** - Transparent brown wings and a light brown body

Management

> » The following measures must be taken 45 days before harvest:
>   - In order to expose the pupa, ploughing must be done in the summer.
>   - Every week, you should get rid of the fallen fruit.
>   - Put in at least four bottle traps for methyl eugenol (0.1% of an acre)
>   - Regularly plough the soil in the tree basin.
> » If you spray your crop with Decamethrin 2.8 EC at 0.5 ml/L and Azadirachtin (0.3%) at 2 ml/L three weeks before harvest, you'll have no trouble picking it off in a timely fashion.
> » For one hour at 48 °C, hot water can be used to cure the collected fruits.
> » Two applications, two weeks apart, of a bait spray containing Malathion 50 EC 2 ml/L or Dimethoate 30 EC 1 ml/L or Carbaryl 50 WP 4 g/L and molasses or jaggery 10 g/L should be applied to the plant two weeks before the fruits are ready to be picked.
> » Natural enemies, like as *Opius* compensates and *Spalangia philippines*, are released in the field.

Stone weevil, *Sternochetus mangiferae*

Damage symptoms

> » Weevils begin laying eggs on developing mangoes when the fruits are the size of marbles.
> » As soon as they hatch, grubs eat through the pulp, tunnel through the seed cover, and enter the seeds, where they pupate.
> » The pulp is drilled through by grubs in a zigzag pattern.
> » Bores into cotyledons and devours immature tissue.
> » At the marble stage, fruit begins to ripen and fall.
> » Bleeding caused by oviposition on fruits the size of marbles.
> » Mature fruit often has tunnelled cotyledons caused by grubs.
> » Weevils emerge with the seed and ruin the pulp right when it's ready to eat.
> » There is just one generation every year, and the adults are dormant for roughly seven to eight months, beginning in August.

Pest identification

» **Grub** - A mature grub has no legs, is fleshy yellow in colour, and has a dark brown or black head.

» **Adult** - The adult stage is characterised by a dark brown body and a rather small nose. After hatching, dark weevils do not actively feed or reproduce; instead, they retreat into the trunk's cracks and fissures.

Management

» Once a week up to harvest, gather up and throw away any dropped fruit.

» Prevent beetles from taking up residence in the bark by spraying it with Chlorpyrifos 20 EC @ 2.5 ml/L just before flowering (November/December)

» When the fruit has reached the lime stage, use Acephate 75 SP at a rate of 1.5 g/L. (2.5 – 4.0 cm diameter). After two or three weeks, you should spray with Decamethrin 2.8 EC @ 1 ml/L.

» It is imperative that any remaining seeds from the orchard or the processing sector be destroyed immediately upon harvest.

» Keep the storage area clean.

Hoppers, *Amritodus atkinsoni, Idioscopus clypealis, I. niveosparsus*

Damage symptoms

» Mango trees attract the insect when they bloom in February.

» Massive numbers of wedge-shaped nymphs and adults can be spotted feeding on the sap of the flower in the spring.

» The infected blossoms wither, become brown, and eventually wither and fall off.

» Once the hoppers have developed, they will move on from the flowers and onto the leaves.

» Sounds like clicking when jassids rustle through the foliage.

» Decline and fall off of flower buds and blossoms.

» Sooty mould, caused by the fungus *Capnodium mangiferae* and *Meliola mangiferae,* reduces photosynthesis in the leaves and lowers the fruit's market value when it covers the inflorescence, the leaves, and the fruits.

» Hopper insects find protection in the nooks and divots of tree bark.

Pest identification

» **Nymph** - Pale yellow nymphs that are quite active and can usually be found hiding in lower shoots or crevices in the bark.

» **Adult**

- *Idioscopus niveoparsus;* dark with wavy lines on wings and three spots on scutellum; adults.

- Adult *I. clypealis* are tiny, a pale brown colour, and have two dark dots on the scutellum.

- Adult *Amirtodus atkinsoni* are sizable, a pale brown colour, and have a pair of dark dots on the scutellum.

Management

» Overcrowded orchards are especially vulnerable, therefore spacing should be kept at a minimum.

» Plowing and weeding are two essential maintenance tasks for orchards.

» Spray Azadirachtin 0.3% at 2 ml/L if the hopper population is low to moderate.

» Spray Imidacloprid 200 SL at 0.25 ml/L or Lambda Cyhalothrin 5% EC at 0.5 ml/L at early panicle emergence if the number of hoppers is more than 4.

» When the fruits are the size of peas, repeat the spraying process if necessary.

» It has been said that the pest can be defeated by using a 1: 1 ratio of Toxaphene and Sulfur.

» Neem oil (at 3% concentration) or Neem extract (at 5% concentration) sprayed.

Shoot borer, *Chlumetia transversa*

Damage symptoms

» In August, the adult moth will lay its eggs on fragile leaves, which will hatch into caterpillars that will bore into the midribs of the leaves.

» After waiting a few days, they tunnel downward into the fragile shoots around the growth point and poop.

» The final shots of the tunnel show it from above looking down.

» Stunting, resulting in terminal bunching of seedlings.

» As a result, the terminal branches die off, the leaves yellow, and the plant withers.

» There is a high mortality rate among newly grafted seedlings.

Pest identification

» **Larva** - The prothoracic shield of young caterpillars is dark brown and th
are a yellowish orange colour.

» Full-grown (20-24 mm) caterpillars are a dark pink with filthy patches.

» **Adult** - The wingspan of an adult moth is roughly 17.5 mm, and its bod
colour is a robust greyish brown with wavy lines. The aft wing membran
are white.

Management

» Lighting traps, pheromone traps, and human intervention are all viab
options for pest management.

» Sanitize your field regularly.

» Plowing throughout the summer to expose the pupae

» Cut down and eliminate infected branches.

» When a new flush appears, spray it with Carbaryl 50 WP at 4 g/L, Quinalph
25 EC at 2 ml/L, Monocrotophos, Fenvalerate, or Cypermethrin.

» The parasitoid *Bracon greeni* has been released.

» *Megaselia chlumetiae* is a parasitoid species that parasitizes shoot bor
caterpillars by depositing eggs on the caterpillar's integument. The new
emerged fly larvae enter the caterpillar and begin feeding on its organs an
tissues. Once the caterpillar dies, the pupation process can begin.

Leaf webber, *Orthaga exvinacea*, O. *eudrusalis*

Damage symptoms

» Moths can lay eggs on leaves or in silken webbings singly or in groups.

» The caterpillars scrape the leaf surface for food once they emerge from
their eggs.

» They soon spin intricate webs to enclose vulnerable young plants for feedin

» Pupation occurs in silken cocoons or in soil, and several caterpillars can b
discovered in a single webbed-up cluster of leaves.

» If the infestation is severe, the development of the panicle may be hampere

» As a result of webbing and eating, leaves become skeletonized and eventual
dry up.

» When a tree is severely infested, it stops producing fruit and takes on a burned appearance with no panicles.

» Since July is peak webber infestation month, prevention and control measures should begin that month. If not, it will have a negative impact on next year›s flush and flowering.

Pest identification

» **Eggs** - The eggs are tiny and a drab greenish tint.

» **Larva** - Having a brown head and prothroacic shield, larvae have a drab green colour overall.

» **Pupae** - White silken cocoons encasing long brown pupae that have remained connected to leaves.

» **Adult** - this moth is brown with white wavy lines across its fore wings.

Management

» Eliminating and destroying diseased shoots through pruning.

» By using a leaf web removing mechanism created by the Central Institute of Subtropical Horticulture in Lucknow, India, we were able to remove the webs and then burn them.

» Lambda cyhalothrin 5 EC (2 ml / L of water) or Quinalphos 25 EC (1.5 ml / L of water) should be sprayed again after 15-20 days if the infestation has not been eradicated.

» Parasiticide Release The genus *Brachymeria*, the genus *Hormius*, and the genus *Pediobius bruchicida*.

Fruit borer, *Deonalis albizonalis*

Damage symptoms

» The larvae of the mango fruit borer are a serious pest because they bore into the immature kernel of the fruit, killing it, starting at around the time the fruit is the size of a marble.

» Moths emerge from the fruits' exit apertures after having pupated as caterpillars within (or occasionally outside).

» Small holes (that double as exits) surrounded by a dark pale brown ring are a telltale sign of damage.

Pest identification

- » **Larvae:** they have a creamy tint and have a brown head. Two black pattern can be seen on the first thoracic segment of the mature larva, and eleve pink bands run dorsally along the second thoracic segment.

- » **Adult:** the size of a wheatgerm, is brown in colour, and has a pointed nos The wings have a bluish-pink metallic sheen and an ashy wood tone. Brush like black hairs on the upper surface of the mesothoracic femur and tibi distinguish men from females.

Management

- » All rotting fruit must be gathered and thrown away.

- » In the event of a severe infestation, spray Endosulfan 35 EC at 2 ml/l beginning at the marble size and following up with Decamethrin 2.8 E(at 1 ml/L two weeks later.

- » Don't use any pesticides in the last two weeks before harvest.

Gall midge, *Erosomyia indica*

Damage symptoms

- » Cecidomyiids, the insects responsible for leaf galls, deposited their eggs o the underside of the leaves.

- » Little raised wart-like galls appear on the upper surface of the leaves when th maggots emerge from their eggs and dig into the leaf tissues below to fee(

- » As a result, the damaged leaves become severely distorted and fall off earl

- » When young plants are infected with leaf galls, growth is stunted.

- » Inflorescences of *E. indica* are bowed and shrivelled after being bitten b midges.

- » Once the larvae have emerged from the galls and moved to the soil, additiona fungal infections might arise through the tiny holes they left behind.

- » Exit holes, when present, are typically found on the underside of the frui close to where it attaches to the central axis of the inflorescence when th fruit is still immature.

Pest identification

- » **Maggots:** a dull yellow

» **Adults:** Females are slightly smaller than males, but both sexes are tiny, yellow flies that resemble mosquitoes. Their wingspan measures between 1.0 and 1.5 mm. The eye facets are round, while the tarsal claws are toothed and spread laterally. It's been established that men have longer antennae than females do. Each male has distal claspers on its lower abdomen.

Management

» The infected leaves must be gathered and thrown away.

» Upkeep of sanitary conditions on the playing field.

» Sticky traps are used to capture the insects.

» Controlling the inflorescence midge, *E. indica*, requires a spray of Fenitrothion 50 EC at a rate of 1 ml/L applied at the emergence of the panicle.

» Apply a spray solution of 2 ml/L of either Dimethoate 30 EC or Methyldemeton 25 EC.

» Reduce the number of pupae by raking the ground in the tree basin.

» To prevent mature *E. indica* from emerging from the soil, it is recommended to flood the root zone of mango trees just before they bloom.

Mealybug, *Drosicha mangiferae*

Damage symptoms

» Mealybug adults set up shop in crevices in the trunk, on tender shoots, and in panicles.

» Silken pouches containing eggs are often deposited in the soil around a tree trunk between April and May, and the young emerge between November and December.

» The nymphs travel up the trees and lay their eggs on the inflorescence, which results in the death of the flowers and ultimately affects the fruit.

» Nymphs and adults sap the plant, weakening it and destroying inflorescence, which leads to fruit falling off.

» The female leaves the tree in April or May to lay her eggs in the ground, where they will remain dormant until December.

» The inflorescence and leaves are drying out.

» Pinkish nymphs and adults of the mealy bug have been spotted on fruit and the stalk.

- » Create honey dew that settles on mango tree leaves, encouraging the growth of sooty mould fungus and limiting the tree's ability to produce food via photosynthesis.

Pest identification

- » Yeasty eggs, in the shape of an oval, and the colour of yellow.
- » Slender female of a rosy color.
- » The adult has a segmented body that is typically covered in wax, has a clearly delineated abdomen, and has fully formed legs and antennae.

Management

- » Mealy bug removal through selective pruning of afflicted plant sections.
- » Ants, which subsist on the mealy bug honeydew, can be used to control the pests. Mealy bugs depend on ants for transportation and protection from predators.
- » In April/May, you can kill the eggs by digging or scraping 15 cm deep around the tree trunks.
- » Midway through December, at a height of 50 cm above ground and just below the juncture of branching, polythene sheets (400 gauge) or sticky bands (sometimes laced with insecticide) are wrapped around the tree stem to prevent nymphs from climbing up the stem.
- » Use jute thread to secure the stem, and then rub some dirt or fruit tree grease down the band's underside.
- » Chlorpyriphos 20 EC 2.5 ml/L may be used if necessary.
- » At a rate of ten insects per tree, we will be releasing the Australian ladybird beetle, *Cryptolaemus montrouzieri*.

Scale insect, *Chloropulvinaria polygonata*

Damage symptoms

- » To feed, nymphs and adult scales feed on the plant's sap from the leaves and other soft parts, weakening the plants.
- » They also produce honeydew, which encourages the growth of sooty mould on leaves and other soft tree parts.
- » Fruit production and tree growth are stunted by a significant infestation of scale.

Pest identification

- » These remain in one place on the plant at all times.
- » They have a dimension of around 2.5 cm and are covered in a white, waxy, hairy, glassy powder.
- » Female adults lack wings while males do.

Favorable conditions

- » During the months of July and August, when high temperatures and humidity coexist, the scale insect population typically peaks.

Management

- » To reduce the number of scales, only use pest-free plant material.
- » Removal and elimination of diseased branches.
- » At 20-day intervals, spray Dimethoate or Monocrotophos.
- » Two predators of C. polygonata ovisacs are the lycaenid parasitoid *Spalgis epius* and the green lacewing *Mallada astur*.
- » Two prominent parasitoids, *Anagyrus pseudococci* and *Promascidia unfascitiventris*, have been documented on scales.
- » Predators such as the midge *Coccodiplosis* sp. and its larvae, as well as the spiders *Cryptolaemus montrouzieri* and *Chilocorus nigritus*, and the shrew-like predator *Scymnus* sp. It is suggested that the ovisac-loving insect, *Cryptolaemus montrouzieri*, be used to manage the scale insect population.

DISEASES

Anthracnose, *Colletotrichum gloeosporoides* (*Glomerella cingulata*)

Symptoms

- » Intense brownish black sunken patches that gradually enlarge and combine into a blighted appearance that causes leaf drop.
- » Brownish-black necrotic lesions occur on twigs, which subsequently coalesce, displaying twig blight and die-back signs.
- » The affected branches will eventually dry out.
- » Early infection of fruit manifests as black necrotic patches that consolidate and eventually cause stunted fruit development and size.

- » Spots become increasingly noticeable on ripe fruits, decreasing the produce value.
- » At this stage of ripening, the fruit's pulp turns hard, cracks, and rots. Fallin infected fruit is a problem.

Favorable conditions

- » We have rain regularly, high humidity (95-97%), and temperatures betwee 24 and 32 °C (the sweet spot is 25 degrees).

Survival and spread

- » Mummified flower buds and flower brackets clung to dry leaves and defoliate stems.
- » Caused by coming into contact with infected fruit while it is being transporte or stored.
- » Conidia in the air facilitate a secondary spread.
- » The fungus' spores (conidia) are spread passively through the splashing rain or irrigation water.

Management

- » Any infected parts, such as leaves, twigs, or fruit, should be gathered an burned.
- » Leaves sprayed with a 1% Bordeaux mixture, 0.2% Captan, 0.1% Thiophanat or 0.25% Mancozeb twice, 15 days apart.
- » Benomyl or Carbendazim, Topsin M or Baycor, diluted to 0.1% and spraye four times, 15 days apart, prior to harvest.
- » Fruits dipped in a solution containing 500 ppm Benomyl and 900 pp Thiobendazole.
- » Fruits are heated at 50–55 °C for 15 minutes in order to ripen them.
- » 15 days before harvest, cover the fruit on the tree with news or brow paper bags.
- » Starting in October, spray 5g/L of *Pseudomonas fluorescens* (Pf 7 strain) c flower branches at 3-week intervals.
- » *Bacillus subtilis* was shown to be relatively efficient, preventing postharve disease in 50% of fruits.

Powdery mildew, *Oidium mangiferae*

Crop losses

- » Serious yield losses of 30–90% are brought on by the disease (Prakash and Srivastava, 1987).

Symptoms

- » It causes damage to the plant's foliage, flowers, panicle stalks, and fruits.
- » When a disease is really bad, the leaves it has affected will drop off.
- » The panicles develop a white or grey fungal growth that eventually turns black, causing the flowers to fall off.
- » The damaged fruits stop developing and may fall off before they reach pea size.
- » The decrease in blooms and their size causes a decrease in harvest.
- » Many abnormally shaped and coloured peanut-stage fruits are rejected.

Favorable conditions

- » Conditions favourable to the propagation of the disease include rain or mist during the day and cooler evenings while the flowers are in bloom.
- » Rapid dispersal is encouraged by conditions of high wind speed for three to four days, high (30°C), low (15°C), and relative humidity (RH) (73–84%), and low (23–26%).

Survival and spread

- » In its resting phase, the fungus persists on afflicted leaves as mycelium and conidia. Some annual or perennial hosts have also been found to support its survival.
- » Airborne conidia cause a secondary infection to spread.

Management

- » The spread of the disease can be slowed by removing and burning affected leaves, stems, and blooms.
- » Sulphur dusting (250-300 mesh) at 0.5 kg/tree to improve plant health.
- » For example, you may use a spray containing either 0.1% Carbendazim, 0.05% Tridemefon, 0.3% Wettable sulphur, or 0.1% Karathane (First spray may be given soon after flowering, followed by 2 more sprays at fortnightly intervals).

» The efficacy of the fungicide can be greatly improved by adding a sticke such as Teepol, at a concentration of 1 ml/L to the fungicidal solution.

» The Mango vs. Resilient species included Totapuri, Neelum, Zardah Totapuri-Khurd, and Janardan Pasand.

Twig blight, *Phoma glomerata*

Symptoms

» Only on older leaves do we see any signs of this disease.

» Mild, uneven, yellow to light brown spots appear randomly on the lea lamina at first.

» A change in color from brown to cinnamon and an increase in irregularit characterise the progression of lesions.

» When a spot has reached its full maturity, it will have a dark perimeter an a dull grey necrotic centre.

» Spots can consolidate into 3.5-10.0 cm patches when infection is sever leading to twig withering and defoliation.

Management

» Having a healthy diet might help you avoid getting sick from the Phom blight.

» An effective treatment was discovered to be spraying with 0.2% Benomy followed 20 days later by 0.3% Copper oxychloride.

Malformation, *Fusarium moniliforme* var. *subglutinans*

Symptoms

» There are three distinct signs: the bunchy top phase, a lack of flowers, an a lack of leaves.

» In the nursery's bunchy top stage, a clump of thickened, little shoots develop eventually bearing a few tiny, undeveloped leaves. Young plants in nurserie are susceptible to severe stunting. Stunted growth means the top will alway look like it's piled on top of everything else.

» In the case of vegetative malformation, young plants have an abnorma number of stunted vegetative branches. The tops of the seedlings are bloatec and they have a bunchy appearance due to the short internodes that ar clustering together to produce bunches of varying sizes.

» Malformed inflorescence is characterised by atypical panicle development.

» Long-lasting black masses form on a malformed head and dry up.

» The secondary stems become a profusion of tiny leaves, giving the broom a witchy air.

» Leads to stunted development and fewer fruits being produced by the plant.

Favorable conditions

» Winter months (December-January) can range from 10 to 15 °C.

» Trees between the ages of 4 and 8 are particularly vulnerable.

Survival and spread

» Fruit in an infected orchard may contain alive spores.

» Transcontinental transmission occurs through the transport of diseased plant parts.

» Within infected orchards, the disease spreads slowly.

Management

» Infected plants need to be removed immediately.

» Disease-free planting medium is utilised.

» Pruning and burning the malformed panicles of flowers and the malformed shoots of plants at the bottom 15–20 cm of otherwise healthy growth can help reduce the overall rate of malformation.

» Malformation rates can be lowered by applying NAA (200 ppm) in the first week of October (before bud differentiation stages), then deblossoming in late December or January.

» Treatment with a spray containing either Carbendazim (0.1%) or Captafol (0.2%).

Grey blight, *Pestalotiopsis mangiferae*

Symptoms

» Tiny brown or yellow dots appear around the leaf's edges and tips.

» As they mature, they grow larger and darker in colour.

» As time passes, the initially white specks, which initially ranged in size from mm to cm, gradually become grey and aggregate into larger grey patches.

- » A few of the lesions' dark edges were slightly elevated.

- » The fungus' many black acervuli grew on grey necrotic regions on mature lesions.

- » The leaf lamina begins to develop brown patches along the edges and tips. As they mature, they become bigger and darker in colour. The acervuli of the fungus, visible as tiny black dots, are responsible for the spots.

Favorable conditions

- » High rates of disease are seen during the monsoon, when temperatures range from 20 to 25°C and the humidity is high.

Survival and spread

- » Spend nearly a year subsisting solely on mango leaves.

- » Conidia carried by the wind are responsible for the spread.

Management

- » Cut off and throw away any contaminated plant pieces.

- » Use a spray of either copper oxychloride (0.25%), mancozeb (1.0%), or the Bordeaux mixture (1.0%).

- » Spraying Wettable sulphur (0.2%) and Zineb (0.2%) on the leaves just before flowering, Carbendazim (0.3%) when the peas are young, and Zineb (0.2%) when the stones are nearly fully developed.

- » Using *Bacillus subtilis* foliar spray.

- » Chausa Cv. is a hardy variety.

Sooty mold, *Meliola mangiferae, Capnodium mangiferae*

Symptoms

- » On the lamina of the leaf, there is a thin, membranous covering that is black like velvet.

- » In extreme circumstances, the mould spreads across the tree's leaves and twigs until they are entirely black.

- » Under dry conditions, the affected leaves curl and shrivel.

- » Black masses of spores cover the surface of the leaves as the fungus grows on the honey dew excreted by insects.

» The formation of a black encrustation inhibits photosynthesis.

» Its attacks during the flowering stage lead to decreased fruit set and, in extreme cases, fruit loss.

» Black mould can also form on the bottoms of fruit.

avorable conditions

» Disease is especially bad in dense old orchards with little sunlight.

» Sooty mould thrives on leaves where there is a high concentration of insects (jassids, aphids, and scale insects), as their sugary secretions provide the perfect medium for its growth.

» The fungus flourished in the high humidity conditions.

urvival and spread

» The principal source of the disease is the leaves.

Ianagement

» Insect and sooty mould control should be tackled at the same time.

» Insecticide Elosal spraying (900 g/450 litres of water) applied every 10-14 days.

» In 15-day intervals, spraying a mixture of Wettasul (0.2%), Metacid (0.1%), and gum Arabica (3%).

» Fruits are bleached by first being submerged in a solution containing 1/4 pound of chloride of lime and 1/4 pound of boric acid per gallon of water, and then being washed.

» Sooty mould resistance has been reported in the mango cv. Alphanso.

» Use systemic pesticides like monocrotophos or methyl demeton to keep pests at bay.

» Put 1 kilogramme of starch or maida per 5 litres of water and spray it. Added 20 litres of water after boiling. Flakes of dried starch are removed alongside the mould.

tem end rot, *Diplodia natalensis*

:rop losses

» Rot not only wastes 4-6% of fruits but also reduces their market value abroad.

Symptoms

» Beginning at the pedicel's base, the epicarp darkens as the process begin:

» A black, round spot will appear on the fruit over the following few hou:
and it will quickly spread in humid environments, eventually covering t
entire fruit and turning it black in about two to three days.

» Brown and mushy pulp characterise sick fruit.

» After about two to three weeks after harvest, decomposition speeds
significantly.

» *Dipolodia* stem end rot is characterised by a predilection for deterioratic
at both ends.

» With the help of pectinolytic and cellulolytic enzymes, the pathoge
promotes soft rot.

Favorable conditions

» Rain, temperatures between 25 and 31°C, and relative humidities (RH)
80% or higher all promote the spread of disease.

Survival and spread

» Molds and fungi can live off of tree debris, such as bark and twigs.

» Rains can carry disease spores far and wide.

Management

» Fruits are submerged in a Borax solution at 43 °C for three minutes.

» Infected branches should be removed and thrown away.

» Carbendazim, Thiophanate Methyl, or Chlorathalonil 0.2 % pre-harve
treatments every 14 days throughout the wet season.

» Fruits should be treated with a 6-8% Borax solution or a 300-400 pp:
Microstatin solution after harvest.

Die-back, *Lasiodiplodia theobromae*

Symptoms

» The disease causes the branches to die back from the top down, especial
in the elder trees, and then the branches and leaves fall off entirely, givir
the trees the appearance of having been scorched by fire.

» In the form of dark lesions on the bark, these progressively worsen and eventually kill the young branches.

» The upper leaves progressively turn brown as their healthy green tint fades, and their margins also roll upward.

» If left unchecked, such leaves may dry and fall off in a month or more, leaving the twigs nearly naked.

» Branches that bleed gum before they die develop fissures.

» When a young plant's graft union is compromised, it often perishes.

Favorable conditions

» Summer's scorching heat weakens mango trees, making them easier targets for the disease.

» Rain, temperatures between 31.5 and 25.9 °C, and relative humidity of around 80% all contribute to the spread of disease.

» Germination of spores is most successful at 30 °C.

Management

» Infected sapwood from scions shouldn't be chosen for multiplication.

» It is crucial that new orchards be protected from the spread of disease at all costs.

» Pruning infected areas promptly is essential, as is the subsequent application of Copper oxychloride or cow dung to the snipped-off ends.

» In order to achieve the entire removal of the pathogen, pruning should be performed in such a way that some healthy section is also removed (7.5 cm below the infection site).

» It's recommended to gather up the diseased branches and burn them.

Bacterial canker, *Xanthomonas campestris* pv. *mangiferaeindicae*

Symptoms

» A wide variety of plant tissues, including leaves, petioles, twigs, stems, and fruits, are susceptible to infection.

» Clusters of tiny water-filled lesions can form on any area of the leaf.

» As they swell, they change colour from brown to black and acquire a halo of yellow light.

» Large necrotic areas, often rough and elevated, can occur when many lesions join together.

» Dead leaves fall off as the disease progresses.

» Young fruits often have water-soaked lesions that eventually turn dark brown or black.

» Fruits that are infected may develop skin fissures, and severely damaged fruits may fall off the tree before they are ready.

» Furthermore, saprophytic fungi are easily invited to invade at the canker stage

Favorable conditions

» Temperatures between 25 and 30 °C and relative humidities of 90 % are ideal for the spread of this disease.

» Infection of fruits is caused by rain that falls when flowers and berries are still developing.

Survival and spread

» Infected tree portions can support the bacterial growth and survival.

» Infected leaves can fall off, which helps decrease cankers, but research has shown that the disease can live for up to eight months in dead leaves.

» It has been proven that mango stones are essential for the survival of the virus.

» It has been shown that the pathogen may also live off of weeds.

» When it rains, diseases spread quickly.

» The introduction of contaminated plants into a new location is the main vector for the disease's propagation.

Management

» Among the precautions mentioned are orchard cleaning and seedling certification.

» Pruning diseased branches and putting up windbreaks around the orchard are both good practises.

» In the field, Agrimycin-100 sprayed at a concentration of 2,000 ppm was effective.

» Spraying with 0.1% Carbendazim once a month was proven to be effective

» Many others have also had success with the Bordeaux mixture, copper oxychloride, and copper hydroxide.

» Effective microorganisms include the fluorescent pseudomonads *Bacillus coagulans, Bacillus subtilis,* and *Bacillus amyloliquifaciens.*

» The Mango vs. It has been stated that the varieties Bombay Green, Jehangir, Fazli, and Suvarnarekha are resistant.

References

Iram, S., Laraib, H., Ahmad, K. S., & Jaffri, S. B. (2019). Sustainable management of Mangifera indica pre-and post-harvest diseases mediated by botanical extracts via foliar and fruit application. Journal of Plant Diseases and Protection, 126(4), 367-372.

Kumar, A., Singh, R., Singh, S. P., Pal, D. S., & Kumar, S. (2020). Economics of different treatments for the management of mango hopper (Amritotus atkinsoni). Journal of Pharmacognosy and Phytochemistry, 9(2), 1729-1731.

Kumar, P., Ashtekar, S., Jayakrishna, S. S., Bharath, K. P., Vanathi, P. T., & Kumar, M. R. (2021, April). Classification of mango leaves infected by fungal disease anthracnose using deep learning. In 2021 5th International Conference on Computing Methodologies and Communication (ICCMC) (pp. 1723-1729). IEEE.

Manikandan, P., Suguna, K., & Saravanaraman, M. (2021). Population dynamics of defoliating insect pests of mango in the coastal agroecosystem of Tamil Nadu. Pest Management in Horticultural Ecosystems, 27(2), 196-200.

Munj, A. Y., Sawant, B. N., Malshe, K. V., Dheware, R. M., & Narangalkar, A. L. (2019). Survey of important foliage pests of mango from South Konkan region of Maharashtra. Journal of Entomology and Zoology Studies, 7(1), 684-686.

Sharma, A., Sharma, I. M., Sharma, M., Sharma, K., & Sharma, A. (2021). Effectiveness of fungal, bacterial and yeast antagonists for management of mango anthracnose (Colletotrichum gloeosporioides). Egyptian Journal of Biological Pest Control, 31, 1-11.

Singh, H. S., & Baradevanal, G. (2021). Mango insect pests and their integrated management strategies. Indian Horticulture, 66(4).

Veling, P. S., Kalelkar, R. S., Ajgaonkar, L. V., Mestry, N. V., & Gawade, N. N. (2019). Mango disease detection by using image processing. International journal for research in applied science and engineering technology, 7(4), 3717-3726.

14

Mulberry

INSECT PESTS

Pink mealy bug, *Maconellicoccus hirsutus*

Damage symptoms

- » Chlorosis (yellowing), deformation (curling), premature drop, stunted growth, and ultimately death of plants are all signs that manifest themselves on the leaves.

- » Adults and juveniles alike cause evaporation by sucking the phloem cell sap from young leaves and buds.

- » The nutritional value of leaves, the amount of leaves produced, and the overall height of the plant are all severely diminished.

- » While eating, the pest injects a harmful chemical into the plants, resulting in dark green, wrinkled, and curled leaves with malformed apical shoots.

- » If the infestation is bad enough, the leaves will become a light yellow colour

- » The affected areas become brittle and fragile.

- » The collection of symptoms is known as "Tukra" (Bushy top) disease.

Pest identification

- » Greenish yellow eggs are deposited in a protective sac.

- » Both the male and female nymph and adult stages are green, while the male pupa and adult stages are pink.

- » The average length of a mature male or female is about 3 mm. Pink in colour females are covered in a white wax and lack wings. Males can fly thanks to their wings and long, waxy tails.

Spread

» Ant traffic aids in the dispersal of mealybugs in the area.

Management

» Snip off infected shoot tips and burn or soak in soapy water to kill any remaining pests.

» Mealy bug populations can be eradicated by removing and burning all crop remains from the affected garden.

» Plants such as hibiscus, beans, pumpkin, croton, chrysanthemum, citrus, grapevine, guava, coffee, sugarcane, soybean, mango, pigeon pea, maize, cotton, and teak, which are all hosts for the mealy bug, should not be grown near mulberry orchards.

» Fish oil rosin soap at 40 g/L or Dichlorvos 76 WSC at 2 mL/L for spraying.

» Spray with Dimethoate (36% EC) at 0.05% 12-15 days after pruning. If you want to keep the pest from returning to your mulberry plants while they're actively growing this summer, you'll need to spray them again with 0.2% DDVP (76% EC) 10 days after the first application.

» Predatory ladybird beetles, *Scymnus coccivora*, and *Cryptolaemus montrouzieri*, will be released at densities of 625 and 1,500 individuals per hectare, respectively.

» The pest-infested hotspot locations will be inoculated with 500 of the exotic Encyrtid parasitoids *Acerophagus papaye*, *Pseudleptomastix mexicana*, and *Anagyrus loecki*; releases may be repeated if necessary.

» Avoiding the use of chemical pesticides can help preserve the released parasitoids and other natural predators like Spalgis and coccinelids.

Leaf webber, *Diaphania pulverulentalis*

Damage symptoms

» Pests that eat the leaves, or defoliators, are a major cause of harm to mulberry trees.

» Silken threads are used by the larva to tie the blades of the mulberry leaf together at the sensitive sprout area, where it then hides and feeds on the leaf's tender green tissues.

» Mature caterpillars devour young leaves, and their excrement can be seen floating in the air below the diseased stalk.

» Because this pest feeds on and damages the growing tip of the shoot, stunts plant development, which has a negative effect on leaf production

» This has a devastating effect on leaf quality and productivity.

Pest identification

» Pinkish, flat eggs.

» The mature, black-headed larva is a greenish brown colour with black pattern on the sides and back of its body segments.

» The pupae of these insects are a rich, dark brown.

» The full-grown specimens have a body length of about 10 mm and a greyis white coloration with brown bands on the forewings.

Management

» Swarm the dried leaves and burn them to destroy the pupae.

» To get rid of adult moths, set up light traps at a rate of one to two trap per acre.

» Put dry sticks all around the garden's edges to attract birds that will e the larvae.

» The pupae are vulnerable to the elements and predators when the soil deeply tilled.

» Pupae can be killed by flooding, which is why it is often used.

» Twelve to fifteen days after pruning or leaf harvest, spray with 0.076 DDVP 76% EC (1 ml/liter).

» From day 5, after chemical spraying, release egg parasitoid *Trichogramm chilonis* at the rate of 1 Tricho card per acre every week for 4 weeks.

» *Bracon brevicornis*, a larval parasitoid, was released in a population of aroun 200 adult wasps.

» Distribute 50,000 *Tetrastichus howardii* parasitoid pupae per hectare. N insecticides should be used in the garden after releasing these parasitoid

Thrips, *Pseudodendrothrips mori*

Damage symptoms

» Both nymphs and adults use their lacerating mouthparts to puncture th epidermis of mulberry leaves and drink the plant sap.

» Saliva secreted during laceration coagulates sap, creating white streaks in the early stage, followed by silvery blotches intermingled with little black patches of thrips faeces.

» The silvery areas on the leaf turn brown and compress as the leaf tissue beneath the epidermis dries.

» Because the leaf tissues have dried out, the leaf curls, and the leaf eventually shrivels and falls off.

» In severe cases, plants stunt, and leaves curl and distort.

Pest identification

» The eggs are a tiny 0.25 mm in length and 0.10 mm in width and are a whitish bean form.

» A first-instar nymph is transparent at first, then turns a pale buttery yellow as it matures; its complex eyes are a deep crimson. On average, a nymph in its second instar is about 0.70 mm long and 0.23 mm wide.

» The pupa has yellow skin and two sets of little wing pads.

» The full-grown adults measure about 0.8 mm in length. Males are a darker tint than females, who are a brighter yellow

Management

» To get rid of thrips, it's important to clean up a mulberry field after harvest by getting rid of weeds, dead leaves, and small branches.

» The pupae of thrips can be killed by the sun and predatory insects if you regularly plough and dig in your mulberry field.

» The use of sprinkler irrigation has been shown beneficial in eradicating both nymphs and adults.

» Frequent watering helps increase pupal mortality in soil, which decreases thrips population growth and subsequent emergence.

» After 15 days of pruning, a 2 ml/liter Dichlorvos 76 WSC spray is used.

» Insecticide application followed by the release of either 500 adult *Scymnus coccivora* or 2500 eggs/ha of Chrysoperla.

Spiraling whitefly, *Aleurodicus dispersus*

Damage symptoms

> » In order to get their nourishment, adults and juveniles use a thin stylet to pierce the leaf and syphon plant sap. This causes the leaf to curl, the plant to develop chlorosis, and the leaves to fall off.

> » An abundance of the powdery wax material excreted by all stages of the insect is easily dispersed via wind, making it a persistent and annoying problem.

> » Honeydew, a sugary secretion of the insects, will collect on the upper surface of the lower leaves and provide ideal conditions for the growth of sooty mould fungus, Capnodium sp.

> » This, in turn, prevents adequate light from reaching the cytochrome tissue of the leaves, so disrupting the photosynthetic process.

Pest identification

> » Eggs have a pale-yellow tint with a short pedicel and are laid with a deposit of a waxy secretion in a spiral pattern.

> » The nymphs in their first instar are active (crawlers), but the ones in their later instars are sedentary.

> » Adult flies are only 2 mm in length and are completely covered in a white powdered wax.

Management

> » Discard the infected foliage and set up 200 yellow sticky traps per hectare

> » Without using chemical insecticides, you can minimise the pest population in the afflicted mulberry garden by spraying it with a powerful jet of water

> » Approximately 15 days after pruning, apply a spray solution of 0.05% Dimethoate 30% EC (1.5 ml/liter), followed by a spray solution of 0.15% DDVP 76% EC (2 ml/liter) applied a week later.

> » A week after applying insecticides, release *Scymnus coccivora* at 1250 adults or *Chrysoperla* at 2500 eggs/ha.

> » In India, *A. dispersus* is known to be parasitized by two species of parasitoids, *Encarsia guadeloupae* and *E. haitiensis*.

> » The discovery of the egg-eating *Axinoscymnus puttarudriahi* has opened the door to eradicating the problem before it even hatches.

Leaf hopper, *Empoasca flavescens*

Damage symptoms

- » Insect nymphs and adults alike cause harm to plants by sucking the sap out of fresh, new growth.
- » Hoppers inject toxins into plant tissues as they sip cell sap.
- » The nutritional value of the afflicted leaves has decreased.
- » Yellow or brown spots on the leaf margins are the first sign, followed by a deformation of the leaf veins.
- » Finally, leaf margins turn dark, dry out, and the entire leaf shrivels up into a cup shape.
- » This painful condition is commonly referred to as "Hopper burn."

Pest identification

- » Adults are roughly 3-4 mm in length and have a pale green or greenish yellow coloration. The wings are braced like a canopy over the abdomen.
- » Adult-looking nymphs don't develop wings until the fourth instar.
- » Eggs are elongated and a pale golden tint.

Management

- » The adult population can be eradicated with the help of light traps and yellow sticky traps.
- » Spraying the area with sprinklers helps keep the bug at bay.
- » Mix 2% rosin soap with 3% neem oil and spray.
- » Strong jets of water sprayed in the afflicted mulberry garden assist reduce the pest population below the level of economic injury.
- » Natural enemies, such as coccinellids and spiders, can be bolstered by growing intercrops like cluster bean, cowpea, black gramme, or groundnut alongside mulberry.
- » Dimethoate 30% EC (3 ml/liter) 0.1% Spray.

Termite/White ant, *Odontotermes obsesus*

Damage symptoms

- » They attack the root systems of fresh mulberry plants and nurseries.

- » They eat tough materials like bark and timber.
- » This causes the cuttings to dry out and fail to sprout.
- » They start their infestations in old plantations in the dry twigs.
- » A while later, they make the gradual transition to live twigs.
- » They produce underground tunnels that stretch from the main stem and are used for foraging.
- » In the case of trimmed plants, they wrap themselves in a protective sheath and consume the twigs.
- » Because of this, they have an effect on the young shoots.

Pest identification

- » The queen has a monstrous stomach.

Management

- » Prior to planting your new mulberry cuttings, make sure you treat them with 0.1% Chlorpyriphos 20% EC solution.
- » Take out the old, dried-out branches and leaves.
- » Termites can be discouraged with the use of flood irrigation.
- » If there are any mulberry gardens in the area, you should seek out the termite mounds and destroy them by breaking them up and killing the queen. It's common for the colony to disband after the queen has been killed or destroyed.
- » Make a Chlorpyriphos 20% EC @ 3 ml/liter solution and pour it into the mound before filling the hole with water and soil.
- » Soil drenching with 0.1% Chlorpyriphos 20% EC should be done in a mature plantation.

DISEASES

Leaf spot, *Cercospora moricola*

Symptoms

- » In the early stages of the disease, little reddish patches with irregular borders form on the leaves.
- » The spots grow, merge, and eventually create shot holes as the severity of the disease increases.

» Yellowing and premature drop of leaves characterises a severe case of this disease.

Management

» A 0.1% solution of the systemic fungicide Bavistin 50 WP was sprayed. When treating a severe case of the condition, two sprays are administered 15 days apart.

» *Myrothecium roridum* can be controlled using a spray of Foltaf 80W 0.2%.

» Mulberry varieties with less than 5% disease incidence include Kanva-2, S-54, MR-2, C-799, Jodhpur, Paraguay, RFS-135, RFS-175 and Almora local. Immune variants include Kalia Kutahi and Bilidevalaya.

Powdery mildew, *Phyllactinia corylea*

Symptoms

» White powdery patches appear first on the underside of leaves, and then spread to cover the entire leaf.

» A disease causes the leaves to transform from green to a range of shades of brown, black, and yellow before dropping off.

» White powdery spots on the underside of the leaves are a telltale sign of this disease.

» As the disease progresses, the blackish-colored patches spread throughout the leaf's entire surface.

Management

» Give the garden more breathing room and light by putting distance between the plants.

» Karathane EC (0.2%) or Bavistin (0.2%) sprays are useful for disease prevention and control. Two sprayings at 15-day intervals may be necessary if the disease is particularly severe. The fungicide spray should completely soak the underside of the leaves.

» Letting out some ladybugs that eat mildew fungus-specifically, yellow and white spotted ladybird beetles.

» Breeding for resistance to mildew. Research shows that Kalia Kutahi is resistant to the disease. Below 5% disease incidence was seen in several cultivars as Mandalaya, Katania, China-White, Jodhpur, Punjab local, MR-2, Almora local, Himachal local, and S-523.

Twig blight, *Fusarium pallidoroseum*

Symptoms

» Infected plants tend to grow thicker and more tangled than healthy one Slight leaf browning/blackening at first, followed by total leaf burn ar then severe defoliation.

» The summer is peak season for this occurrence.

» At first, only individual leaf blades begin to wilt, and subsequently the disea spreads throughout the entire plant.

» Longitudinal black lesions characterise affected branches; these diseas eventually cause the wood to break and dry out.

Spread

» Disseminated by the ground and water.

Management

» Basin irrigation can help stop the transfer of disease from an infected pla to a healthy one.

» Dig up dead vegetation.

» A 20-ton-per-hectare application of FYM.

» Copper oxychloride should be poured at the root's surface at a rate of 2 g per litre of water.

» Foltaf 80W and Dithane M-45 were the most effective fungicides again twig blight among those tried.

» The fungicides can be applied as a foliar spray at a lower concentratio (0.2%) or as a soil drenching at a greater concentration (0.5%) due to tl disease's dual modes of transmission.

» 25 gm of *Trichoderma viride* and *Bacillus subtilis* should be applied per pla either before or after they are grown.

Leaf rust, *Cerotelium fici*

Symptoms

» The eruptive lesions/spots, about the size of pinheads, range in colour fror brown to black and are caused by a pathogen.

» The disease causes the leaves to turn yellow.

» The leaves prematurely drop off as the severity of the disease increases.

Management

» The most effective way to lessen the spread of the disease is to increase the distance between buildings.

» Fungicides such as Kavach 75 WP and Foltaf 80W, when sprayed at a concentration of 0.2%, are effective in preventing and controlling the disease.

» The disease incidence rate was less than 20% in the mulberry cultivars AB x Phill, K2 x Kosen, ACC-115, and Almora local.

Bacterial blight, *Pseudomonas mori*

Crop losses

» It's a major problem in India, reducing leaf production by 5-10% during the wet season.

Symptoms

» Stunted growth and general twig death.

» Water-soaked brown or black dots with a yellow halo appear on the undersides of leaves.

» Some distortion and wrinkling of the leaves may occur.

» When the disease is really bad, the leaves curl, decay, and turn a dark brown or black.

» Infected juvenile shoots become elongated and cankered.

Management

» Remove dead branches in the fall and burn them.

» When blighted shoots are spotted in the spring, they should be immediately lopped off and burned.

» It's best to burn the affected plant after it's been uprooted.

» The polluted ground needs to be left exposed in the sun to dry.

» It is possible to employ a foliar spray with 0.1% Streptomycin or 0.1% Streptocycline to combat bacterial infections.

References

Chandrasekharan, K., & Nataraju, B. (2008). Studies on white muscardine disease of mulberry silkworm Bombyx mori L. in India-a review. *Indian Journal of Sericulture*, *47*(2), 136-154.

Datta, S. C. (2006). Effects of Cina on root-knot disease of mulberry. *Homeopathy*, *95*(02), 98-102.

Kumari, V. N. (2014). Ecofriendly technologies for disease and pest management in mulberry-A review. *IOSR Journal of Agriculture and Veterinary Science*, *7*(2), 1-6

Mahalingam, C. A., Suresh, S., Subramanian, S., Murugesh, K. A., Mohanraj, P., & Shanmugam, R. (2010). Papaya mealybug, Paracoccus marginatus-a new pest on mulberry, Morus spp. *Karnataka Journal of Agricultural Sciences*, *23*(1), 182-183

Misra, C. S. (1920). Tukra disease of mulberry. In *Report of the Proceedings of the Third Entomological Meeting held at Pusa on 3rd to 15th February 1919*. (pp. 610-618) Superintendant Government Printing.

Narayanaswamy, K. C., Geethabai, M., & Raghuraman, R. (1996). Mite pests of mulberry-a review. *Indian Journal of Sericulture*, *35*(1), 1-8.

Rahmathulla, V. K., Sathyanarayana, K., & Angadi, B. S. (2015). Influence of abiotic factors on population dynamics of major insect pests of mulberry. *Pakistan Journal of Biological Sciences*, *18*(5), 215-223.

Sharma, A., Sharma, P., Thakur, J., Murali, S., & Bali, K. (2020). Viral diseases of Mulberry Silkworm, Bombyx mori L.-A Review. *Journal of Pharmacognosy and Phytochemistry*, *9*(2S), 415-423.

Sharma, B., & Tara, J. S. (1985). Insect pests of mulberry plants (Morus sp.) in Jammu region of Jammu and Kashmir state. *Indian Journal of Sericulture*, *24*(1), 7-11

Sharma, D. D., Naik, V. N., Chowdary, N. B., & Mala, V. R. (2003). Soilborne diseases of mulberry and their management. *International Journal of Industrial Entomology*, *7*(2), 93-106.

15

Papaya

INSECT PESTS

Mealy bugs, *Phaenacoccus marginatus*

Damage symptoms

- » Pesters plant life and edibles
- » Leaves are turning yellow.
- » Injection of a poison causes malformation in the affected area.
- » Slower than normal leaf development, together with a decrease in fruit production.
- » Red and black ants are present.
- » *Capnodium* fungus causes honey dew and sooty mould growth.
- » Complete annihilation of the plant

Pest identification

- » **Egg:** In an ovisac that's three to four times the length of the body and completely covered with white wax, the female lays eggs that are greenish yellow in colour.
- » **Nymph:** Crawlers or nymphs, to be precise.
- » **Adult:** the female has a bright yellow colour and is around 3 mm in length and 1.4 mm in width; she also has a thin, white waxy coating. Male adults measure about 1 mm in length and are a bright pink colour.

Management

» Removal and destruction of diseased branches through pruning.

» Clearing and combustion of crop waste.

» Suppressing ant populations and eradicating established anthills.

» Crawler movement can be halted with the use of sticky bands, an alkathen sheet, or an insecticide band applied to the branches or main stem.

» Neem oil (1–2%), NSKE (5%), and fish oil rosin soap (25g/liter of water) are all examples of botanical pesticides that can be used.

» Seedlings are treated for mealy bugs with a one-hour soak in Dichlorvo (76% EC at 10 ml/L of water) before planting.

» For severe infestations, use insecticides at the appropriate rates, such a Profenophos 50 EC (2 ml/L), Chlorpyriphos 20 EC (2 ml/L), Buprofezi 25 EC (2 ml/L), Dimethoate 30 EC (2 ml/L), Thiomethoxam 25 WG (0.(g/L), and Imidacloprid 17.8 SL (0.6 ml/L).

» Saturating the soil with Chlorpyriphos around the plant's collar will dete mealy insect and ant activity.

» There are many predators of the papaya mealy bug, including the *Cryptolaemu montrouzieri, Spalgis epius, Rodolia fumida,* and *Scymnus* sp.

» Parasitoids like the exotic *Anagyrus loecki, Acerophagous papayae, Phygadiun* spp., and *Pseudleptomastrix mexicana* are just a few examples (Hymenoptera Encyrtidae).

» Entomopathogenic fungus *Beauveria bassiana* is a parasite of mealybugs.

Whitefly, *Bemisia tabaci*

Damage symptoms

» White nymphs feed on sap from the undersides of leaves.

» Damaged leaves become yellow and curl downward.

» They cause premature shedding and wrinkles.

» Papaya leaf curl disease is spread by these whiteflies.

Pest identification

» A pear-shaped egg is placed on the underside of a leaf.

» Soon after hatching, the nymphs find the same leaf and begin feeding there

» Adults are little flies with a white waxy bloom on their bodies.

» In a year, there are roughly 12 generations.

Host range

» Tomatoes, tobacco, cotton, and even some winter veggies are all vulnerable.

Management

» Yellow sticky traps are set up.

» At the first sign of an infestation, use a spray containing either Imidacloprid 200SL at 0.01% or Triazophos 40EC at 0.06%.

» The lambda cyhalothrin 5 EC at 0.5 ml/L or the decamethrin 2.8 EC at 1 ml/L should be sprayed.

» Try NSKE 5% or 3% Neem oil spray.

» Predators, including the *Cryptolaemus montrouzieri* and the *Dicyphus hesperus*, are set free.

» Parasitoids such as the *Encarsia haitiensis, Encarsia guadeloupei, Encarsia formosa, Eretmocerus species*, and *Chrysocharis pentheus* were let loose.

Fruit fly, *Bactrocera (Dacus) dorsalis*

Damage symptoms

» Females use their pointed ovipositor to pierce the outer wall of ripe fruit, where they deposit their eggs in clusters within the mesocarp.

» Once maggots hatch, they feed on the fruit pulp, causing the infected fruit to decay.

» Fluid seepage and dark, rotting spots on fruit.

» Fruits falling from the sky.

Pest identification

» **Egg:** White, elongated, and roughly the size of a quarter of an inch, eggs have a half an inch on each side.

» **Larva:** a cone with no legs. The body tapers to a point at the mouth. Third-stage instars are roughly half an inch in length.

» **Pupa:** a seed and is a yellowish brown colour.

» **Adult:** the abdomen has three distinct black stripes: two horizontal ones at the navel and a long one running from the third segment's base to th apex. Often, but not always, these marks are in the shape of a "T," howev the exact shape can be very difficult to determine.

Management

» Collect the rotten fruit and bury it in a pit to get rid of the pests.

» Pupae can be exposed with the help of summer ploughing.

» Fly activity can be tracked with methyl eugenol sex lure traps.

» Apply a solution of Fenthion 100 EC 2 ml/ L or Malathion 50 EC 2 m L by spraying it.

» Dispersal of Predators and Prey *Spalangia philippines, Diachasmimorp kraussi, Fopius arisanus, Opius compensates* and *O. fletcheri.*

» Ants prey on fruit flies.

Green peach aphid, *Myzus persicae*

Damage symptoms

» Sap is sucked from leaves, petioles, and fruits by nymphs and adults alik

» Having its leaves curl and fall.

» Plant growth retardation

» Excreta of honeydew promotes the growth of black sooty mould.

» Early ripening fruit failure.

Pest identification

» **Egg:** small, glossy black objects.

» **Nymph:** Immature aphids, known as nymphs, resemble full-grown aphic in appearance but are much smaller.

» **Adult:** tiny (1–4 mm in length) horned insects with two long antenna Most species of aphids are equipped with two diminutive cornicles (horn at the insect's posterior end.

Management

» Take out the broken pieces and throw them away.

» Use a spray containing 0.03% dimethoate or 0.025% methyl demeton.

» Spreading parasitoids like *Aphelinus mali* and *Aphidius colemani* in the wil

» *Coccinella septumpunctata*, a natural predator, has been released into the wild.

Leaf hopper, *Empoasca stevensi*

Damage symptoms

» Hopper burn, the result of leaf hoppers' feeding, is a dry, shrivelled appearance of the affected leaves.

» Saliva is harmful to plants, causing yellowing, curling, drying, and tissue death in the leaves.

» In other words, the plant's growth is stunted.

» Bloody white latex seeps out of the plant where it has been pierced, typically along the leaf veins and the petiole.

Management

» Imidachloprid or Malathion spray.

White peach scale, *Pseudaulacaspis pentagona*

Damage symptoms

» White peach scale is a "triple menace" to farmers since it can infest not only the bark but also the fruits and leaves of plants.

Management

» White peach scale is prey for several predators, including ladybird beetles and common lacewings.

» White peach scale is also a target for some gall midges.

» The parasitoid wasp Encarsia diasapidicola is being researched as a potential method of controlling white peach scaling.

Red spider mite, *Tetranychus cinnabarinus*

Damage symptoms

» Spider mites use their long, needle-like mouthparts to eat the cellular contents of leaves. This causes the chlorophyll content of the leaves to decrease, which in turn causes the leaves to develop white or yellow spots.

» The underside of the leaves may have clusters of yellow spots, while the upper surface may be covered with silken webs spun by tiny red mites.

» When an infestation is bad enough, the leaves wither and fall off.

» Under adverse conditions, the mites also generate webbing on the lea surfaces.

» In dense populations, the mites "balloon," or gather into a sphere of sil threads that is then carried by the wind to other parts of the plant.

Pest identification

» **Egg:** round, reddish in colour, and have a little filament protruding fror one end.

» **Nymph:** The nymphal stages of development include the six-legged larv the protonymph, and the deutonymph.

» **Adult:** Female adults are elliptical in shape, with a brilliant crimson fron and a dark purplish brown back.

Management

» Dicofol EC 18% @ 2.5 ml/l spray.

» For example, the Anthocorid insect (*Orius* spp.), the green lacewings Mallad basalis and *Chrysoperla* sp., the predatory mites *Amblyseius alstoniae, A womersleyi, A. fallacies,* and *Phytoseiulus persimilis,* the predatory coccinelli (*Stethorus punctillum*), the staphylinid be (*Feltiella minuta*).

Broad mite, *Polyphagotarsonemus latus*

Damage symptoms

» Young leaves and their growth points are attacked, hardening and distortin them.

» The leaves turn a sickly yellow and develop sharp, claw-like margins.

» Scar tissue between veins on the underside of leaves turns grey or golden

» Fruits are another target.

Management

» Apply Dicofol twice, 10-14 days apart.

» Sufficient protection is necessary.

DISEASES

Anthracnose, *Gloeosporium papayae*

Symptoms

- » Initially appearing as little, dark brown discolorations on the surface of fruits, the spots eventually grow to become larger, sunken circles anywhere from 1 to 3 mm in diameter.

- » Spots begin to merge and generate a thin layer of mycelium around their edges.

- » In damp environments, a salmon pink encrustation of spores is released.

- » Fruits that are infected at a young age will shrivel and turn mummy-like if left untreated.

- » The leaves and stems also develop necrotic patches.

- » Petioles with concentric rings of acervuli indicate where they have recently erupted.

- » Small lesions are caused by the pathogen even when the fruit is still in its green stage, although it typically manifests when the fruit is ripening.

- » As long as the fruit is still green, lesions grow very slowly, and rarely reach a diameter of more than 12 mm.

Spread

- » It has been scientifically proven that field fruit can spread infection.

- » Splashes of rain facilitate the secondary spread of conidia.

Management

- » Collecting diseased leaves and removing them.

- » Fruits are heated in water at 45 °C for 30 minutes.

- » Every 10-14 days, spray 0.25% Mancozeb, 0.1% Thiophanate methyl, or 0.2% Chlorothalanil.

- » Using 0.1% Carbendazim every 45 days or 0.2% Daconil every 15 days, two sprays were successful.

Powdery mildew, *Oidium caricae*

Symptoms

» White patches caused by fungi can be found on the underside of leaves. These patches use haustoria to absorb nutrients from the leaf's cells.

» Pale yellow or green spots can be seen on the upper surface of leaves, typically along veins where infections have occurred.

» Powdery growth is a common symptom of fruit infections caused by this pathogen.

» Leaves become yellow and fall off completely when attacked severely.

Pathogen identification

» An obligatory parasite, really.

» Hyaline, septate mycelium with haustoria arising from epidermal cells characterise this fungus.

» Its conidia have a hyaline structure.

Management

» A spray regimen of 0.3% wettable sulphur applied every 10 days is very successful at preventing the spread of disease.

» Spraying systemic fungicides at 0.1% concentration once a month was more successful than using spot treatments.

Phytophthora blight, *Phytophthora nicotianae* var. *parasitica*

Symptoms

» The fruiting or leaf scarring regions of the stem are the primary targets of the pathogen.

» The infected region becomes larger and finally girdles the stem of the young tree, causing the top of the plant to wilt and die.

» While still on the tree, fruit is susceptible to infection at any time.

» As it dries out, fruit becomes brown and eventually falls to the ground.

Management

» It is important that the papaya stems not be injured or damaged in any way.

» It is crucial that sick trees and fruit be promptly removed from the orchard and disposed of in a safe manner.

» For effective control of the disease's airborne transmission, spraying with Dithane M-45, Foltaf, Ridomil, Aliette, or Chlorothalanil at a concentration of 0.2% every two weeks is recommended.

» The papaya 'Kapoho Solo' variety is the disease-resistant one.

Foot rot, *Pythium aphanidermatum*

Symptoms

» Symptoms include the development of water-soaked areas on the stem at or near the ground.

» Rapid growth and girdling of these areas around the stem indicates rotting tissues, which ultimately turn a dark brown or black colour.

» Poorly rooted plants are unable to endure high winds and sometimes collapse and die as a result.

» If the disease isn't very severe, only one side of the stem will rot, and the plants will stay small.

» If any fruit forms at all, it is little and misshapen.

» The plant progressively withers and dies.

Favorable conditions

» When it rains and the humidity is high, diseases can spread rapidly in already weakened soil.

Survival and spread

» The oospores germinate and release the zoospores, which are carried by the irrigation water and eventually spread across the field.

Management

» At the time of planting, apply a mixture of *Trichoderma viride* (15 g/plant) and well-decomposed FYM to the soil surrounding the plants' root zones.

» To avoid water damage to the crop's stems, the ring method of irrigation should be used when watering the crop.

» If the soil doesn't become waterlogged before a new planting, the disease might be contained.

» Two to three litres of Copper Oxychloride should be soaked into the soil (3 g per litre of water). Mancozeb (2.5 gm per litre of water) is another option.

» In the event of a disease attack on already established crops, scrape th rotten area of the plant and apply Copper Oxychloride or Bordeaux past

» During fruit development, irrigate the plant twice at 15-day intervals wit Mancozeb (2.5 g/liter of water).

Ring spot virus disease, *Papaya Ring Spot Virus* (Potyvirus)

Symptoms

» Symptoms consist of the upper surfaces of the terminal leaves clearin puckering, or bulging in the tissue between the secondary veins and veinle

» Young leaves often curl inward and downward at the edges and tips.

» Plants infected with this virus exhibit a variety of symptoms, includin mosaicing, mottling, dark green blisters, necrosis of chlorotic patches, le distortion leading to shoe string symptoms, and ultimately, stunting.

» The stems of immature plants exhibit mosaic or mottle symptoms, as we as dark green patches and oily or water-soaked streaks.

» Smaller, heavily lobed, and more sideways than round, the fruits exhib concentric ring patterns.

» The sugar content in diseased fruits drops by 40%, and the latex quali declines.

Spread

» Aphids and mechanical means both contribute to the spread of the viru *Myzus persicae* and *Aphis gossypii* are the most effective vectors.

Management

» Protect papaya seedlings from pests by growing them in a nursery.

» Spread healthy seedlings.

» Before planting papaya, it is recommended to grow sorghum or maize a a barrier crop.

» Upon seeing the signs, remove the infected plants immediately.

» Aphid vectors can be captured with yellow sticky traps set along the border

» Avoid growing pumpkins and squash near the playing field.

» Nimbecidine 0.4% or 20 ml/L of groundnut oil can be sprayed.

» Comparing papaya with something else. The ring spot virus cannot infe Rainbow and Sunup.

Mosaic disease, *Papaya Mosaic Virus*

Symptoms

- » Papaya leaves become mosaic and the plant itself becomes stunted as a result.
- » About 5 days after being inoculated, young seedlings in the greenhouse begin to display vein-clearing and downward cupping of the leaves.
- » After about two weeks, a mottle or mosaic appears.
- » New leaves are where symptoms first show up on plants.
- » The leaves are abnormally small and blistered looking, with dark green tissue in between the lighter green lamina.
- » The length of the leaf petiole is shortened, and the upper leaves stand erect.

Spread

- » Aphids are the vectors for this disease.

Management

- » Eliminating infected plants and destroying them is an important part of practising good field sanitation.
- » Preventing aphid infestations is essential for limiting financial losses.
- » Aphid populations can be efficiently controlled by using Carbofuran (1 kg a.i. /ha) at the time of sowing seeds, and then applying Phosphamidon (0.05%) as foliar sprays three times, beginning 10 days after sowing and continuing for a total of three applications.

Leaf curl disease, *Papaya Leaf Curl Virus*

Symptoms

- » Leaves get distorted and curled, their lamina thin, their margins roll inward and downward, and their veins thicken.
- » Leathery, brittle, and twisted leaves are the result.
- » Vegetation shrank and failed to flourish.
- » Damaged plants will not bloom or bear fruit.
- » These symptoms can sometimes be seen on all of the upper leaves of the plant. The plant's growth stops and its leaves die off when the disease has progressed far enough to do so.

» *Bemisia tabaci*, often known as the silver leaf whitefly, is the insect responsibl
for spreading the virus.

Spread

» The virus can easily spread via grafting and the white fly, Bemisia tabaci.

Management

» Only by eradicating and disposing of infected plants can we hope to sten
the disease's spread.

» Decreased infection severity can also be achieved by controlling white fl
populations.

» Whiteflies can be effectively managed by applying Carbofuran (1 kg a.i./ha) t
the soil at planting time, followed by four or five foliar sprays of either 0.05%
Dimethoate, 0.02% Metasystox, or 0.05% Nuvacron at 10-day intervals.

References

Biratu, W. (2022). Papaya Fruit Pests and Development of Integrated Pes
Managements: Critical Review. *Journal of Biology, Agriculture and Healthcar*
12, 15.

Kalyanasundaram, M., & Mani, M. (2022). Pests and Their Management o
Papaya. *Trends in Horticultural Entomology*, 671-688.

Meena, M. B., & Singh, M. K. Integrated Disease and Pest Management Strategie:
A Holistic Approach for the Sustainable Production of Papaya. *Just Agricultur*
Multidisciplinary e- Newsletter, 3, 3.

Mishra, P. P., & Sasmal, A. (2020). Recent Advances in the Management of Mealybug
with Special Reference to Papaya Mealybug in India. *Recent Trends in Insec*
pest management, 115, 133.

Pacheco-Esteva, M. C., Soto-Castro, D., Vásquez-Lopez, A., Lima, N. B., an
Tovar-Pedraza J. M. (2022). First report of *Colletotrichum chrysophilum* causin
papaya anthracnose in Mexico. *Plant Disease, 106*(12), 3213.

Persley, D. M. and Ploetz, R. C. (2003). Diseases of papaya. *Diseases of tropical frui*
crops, 373-412.

Prajapati, B. K., & Gohel N. M. (2022). Important Diseases of Papaya (*Caric*
Papaya L.) and Their Management. *Diseases of Horticultural Crops: Diagnosi*
and Management: Volume 1: Fruit Crops, 123.

16

Peach

INSECT PESTS

Plum curculio, *Conotrachelus nenuphar*

Damage symptoms

» Peaches, plums, other stone fruits, apples, and even pears can all be damaged by these pests, both as adults and as grubs.

» The female adult causes most of the damage when she cuts a crescent shape into the fruit's skin to deposit her eggs.

» D-shaped scars form on the surface of the fruit as a result.

» The fruit is wasted when grubs that hatch from the eggs consume it.

» Later in the season, both sexes may puncture the fruit with small, spherical beaks while feeding.

Pest identification

» **Grubs:** Grubs are smooth-bodied, legless, and up to half an inch in length. Its body is a muted grey to yellow, and its brown head is somewhat curled.

» **Adult:** The adult beetle is a rough and warty brown with a mottled appearance. Its snout is about an inch and a quarter in length and curled.

Survival and spread

» Adult plum curculios overwinter in sheltered areas close to orchards, such as under leaves, bushes, or other debris.

Management

- » Proper sanitation measures include removing or cleaning up overwintering sites and picking up and destroying early-dropping fruit once a week.

- » Malathion sprays used in the middle of June, at the end of June, and at the beginning of July will be sufficient to eradicate the second-generation adults.

- » Plum curculio can also be managed with pesticides like lambda-cyhalothrin (Permethrin) or cypermethrin (Permethrin).

- » Use a combination of Neem oil and pyrethrins.

- » Clemson Fruit Bags are a specialised enclosing bag that are removed at harvest time and are ideal for containing green fruit that is about the size of a fingernail (about 3 weeks after bloom).

Oriental fruit moth, *Grapholita molesta*

Damage symptoms

- » Young caterpillars can be seen eating holes into the tips of peach tree branches.

- » As a result of this action, the tips of the branches will begin to shrivel and die.

- » At the end of the season, when the branch tips have hardened, caterpillars gain access and feast on the fruit instead.

- » Caterpillars are safe from insecticides while they are inside the fruit or twigs.

Pest identification

- » **Caterpillar:** The average inch-and-a-half-long caterpillar is pinkish with a brown head and has six distinct legs.

- » **Adult:** This moth's adult form has a drab brown colour with a wing span of only around an inch.

Survival and spread

- » Cocoons made by these pests are hidden in safe spots on the tree or among debris at the tree's base, where the developed larvae can spend the winter.

Management

- » Pheromone-filled traps can be used to detect the presence of moths (synthetic insect attractants).

- » A spray of Permethrin, Lambda cyhalothrin, or Malathion should be used.

if more than 10 moths are caught in each trap, on average.

Tree borer, *Synanthedon exitiosa*

Damage symptoms

» Peach, cherry, and plum trees are all susceptible to assaults by the peach tree borer larva, which can be found eating anywhere from the tree's main roots to roughly 10 inches up the tree's trunk.

» Gum in large quantities, often combined with frass (the sawdust-like excrement of insects).

» An increased number of larvae is tolerable only on older trees.

Pest identification

» **Larva (immature stage):** The full-grown length of a larva (immature stage) is about 1 to 114 inches. The creature's body is a milky white, while its head is a chocolate brown.

» **Adults:** They emerge as adults and are known as clearwing moths. The female borer is primarily metallic blue-black in appearance, with the exception of a red-orange stripe across the abdomen. Males are distinguished by their all-black plumage, which is accented by tiny yellow stripes on the belly and broad yellow stripes along the back, just below the wings.

Survival and spread

» Larvae of the peach tree borer survive the winter.

Management

» To prevent peach tree borer infestations, spray the trunks with Permethrin once a year in August. This is also the best time to spray the trees' bases.

» Be sure to administer sufficient spray from the scaffold limbs to ground level so the bark is soaked and a small puddle formed at the base of each tree.

» Peach tree borer does the most harm to young trees, so it's important to prevent injuring the bark when planting.

» It is highly recommended that you soak your plants in an insecticide solution before you put them in the ground.

» The tree may shift enough on sandy or other loose soils for the wind to erode the bark or create a breach between the trunk and the ground. The larvae will have little trouble getting in through this opening.

» Try poking, mashing, or digging the grubs out with a thin wire.

Lesser tree borer, *Synanthedon pictipes*

Damage symptoms

» The lesser peach tree borer eats the trunk and primary branches of peac cherry, and plum trees.

» Once again, symptoms include frass-filled gum that is seeping.

» When the infestation is severe, it can take off entire trees or even just certa branches.

Pest identification

» **Larva:** The caterpillar looks a lot like the peach tree borer larva, excep smaller.

» **Adults:** They emerge as adults and are known as clearwing moths. Adu male and female lesser peach tree borers look similar to the larger ma peach tree borer.

Survival and spread

» The larvae of the lesser peach tree borer survive the winter.

Management

» Lesser peach tree borer can be prevented by maintaining a healthy growir environment and avoiding any kind of mechanical damage to the trees.

» Remove any branches that are diseased, rotten, or have symptoms of bor damage.

» Destroy the cut branches by shredding or burning them before the adul appear.

» If you're concerned about spreading bacterial canker when pruning, yc can sterilise your tool between cuts by dipping it into a solution of one pa household bleach to nine parts water.

» The lesser peach tree borer can be managed by spraying the tree's trunk an branches with Permethrin in August.

White peach scale, *Pseudaulacaspis pentagona*

Damage symptoms

» White peach scale can quickly weaken, destroy, and defoliate peach trees and their branches.

» The females pierce the leaf or stem with their mouthparts and sucking on the sap.

» Peach trees can be infected on multiple levels, including the bark, the fruit, and the leaves.

» The trees become stunted, lose their leaves, and sometimes even die from the inside out as a result of the infestation.

Pest identification

» **Nymphs:** Nymphs are a juvenile stage that resembles full-grown creatures but are much smaller.

» **Adults:** Teens and adults are similarly little and unable to move. Waxy coatings produced by females are often resistant to pesticides. Females range in size from 1/16 to 1/8 of an inch and have a greyish white, yellowish patch on a circular body.

Survival and spread

» The female white peach scale can overwinter in a warm environment.

Management

» When temperatures are above average during the dormant season (before bud break), spray the trunk and branches thoroughly with 2% horticultural oil to kill any overwintering adult females. It's preferable to apply the spray twice, once at the 3-week mark and once a week before the buds start to swell.

» Malathion, Permethrin, Lambda cyhalothrin, and Cypermethrin are all chemical pesticides that can be used to control only the crawlers.

DISEASES

Brown rot, *Monilinia fructicola*

Symptoms

» Throughout the growing season, this pathogen targets numerous peach and plum plant parts, including flowers, branches, young fruit, and mature fruit.

» On young shoots, brown, sunken patches emerge, and infected branches leaves get wilted and fall off.

» Flower petals could turn brown or wilt.

» The fungus forms a brown spot that spreads and rots on the fruit.

» Fruit that has been infected will begin to rot in small, rapidly expanding patches; the fruit will then shrivel and be coated in a mat of fungal growth that generates fuzzy tan/grey spores; and the spores will stick to the surface of the fruit.

» The infected fruit falls to the ground, and if it is left on the tree, it will dry out, turn black, and solidify into "mummies."

Favorable conditions

» Conditions favourable to brown rot include high humidity, precipitation and warm temperatures.

Survival and spread

» Cankers in twigs, fruit mummies, and peduncles (the stemlike structures that join the flower or fruit to the branch) are all places where fungi can overwinter.

» *Mummies* are fungus spores that have been released from infected fruit.

» The spores are spread by the wind and the rain.

Management

» Proper tree pruning will increase breezes and hasten the drying of fruit and leaves.

» Chlorothalonil-based fungicides are highly effective, but they need to be sprayed before the shuck-split stage for maximum effectiveness.

» Apply fungicide treatments containing Captan or Sulfur after shuck-split to prevent brown rot.

» In a well-drained area, plant a resistant cultivar such as Venture.

Scab, *Cladosporium carpophilum*

Symptoms

» Disease symptoms appear as velvety, olive-green patches on fruit, leaves or twigs.

» The spots begin at a size of about one-eighth of an inch and grow to about that size.

» About 3 weeks after the petals have fallen, you'll see these discolorations.

» When spots do appear on fruit, they tend to be located closest to the stem.

» Many simultaneous infections might cause the fruit to crack.

» The blemishes on the fruit are just on the outside, not the inside.

» Sunken lesions (abnormal tissue growths) may form in the face of a severe disease.

Favorable conditions

» The disease thrives in cool, rainy climates (in spring).

» Germinating spores need 24 hours of high relative humidity.

» Infection rates are higher in regions that receive more precipitation.

Survival and spread

» Infected leaves and twigs serve as winter habitats for the scab pathogen.

» The majority of spores will originate from these wounds.

» There are spores in the air and in the water.

Management

» Take out the cuttings and throw them away.

» In order to prevent overwintering fungus, fall cleanup is mandatory.

» Chlorothalonil-based fungicides are highly effective, but they need to be sprayed before the shuck-split stage for maximum effectiveness.

» Apply fungicide treatments containing Captan or Sulfur after shuck-split to prevent brown rot.

Leaf curl, *Taphrina deformans*

Symptoms

» This disease is one of a kind since the fungus develops through two distinct phases.

» Curled (infected) leaves are a source of one kind of spore that can be found in the spring.

» Both sides of the leaf are susceptible to infection by the fungus.

» Thicker yellow or red patches appear on infected leaves as the fungus spreads

» Because the diseased, thickened region of the leaf is growing at a slower rate than the remainder of the leaf, the leaf curls as it matures.

» These dense regions release spores, which, once germinated, give rise to a new stage of the fungus that develops on and along the growing tips of the shoots.

» Fruits at all stages of development are susceptible to infection.

Favorable conditions

» Wet weather during early bud development is ideal for the disease. Humidity levels higher than 98% are required.

Survival and spread

» Pathogens that cause disease can survive the winter in sheltered places like cracks in the bark or surrounding new growth.

» There are millions of fresh spores produced by the fungus, and they are spread from tree to tree via splashing or wind.

Management

» It is recommended that contaminated fruits and leaves be picked up and disposed of after they have fallen to the ground.

» Late season fruit thinning.

» If you prune your tree in the fall, before you apply any fungicides, you can limit the number of spores that overwinter on the tree, hence reducing the amount of fungicide you need to use.

» The most common methods of control are the use of resistant cultivars (such as Frost, Indian Free, Muir, Q-1-8, and white-fleshed Peach 'Benedicte', and fungicides (such as Bordeaux combination or Copper sprays).

» Spraying with Chlorothalonil, Bordeaux mixture, Lime sulphur, or Copper during the tree's dormant period and again later in the season are both vital

» Once a disease has spread over a stand of trees, not even drastic measures like cutting can stop it.

Rust, *Tranzschelia pruni-spinosae*

Symptoms

- » Clouds of yellow dots form on the upper limb.
- » At the end of the vegetative phase, the black spores (resistant) develop on the lower right, below the dots.
- » Fungus attacks cause limited frost tolerance and only produce fruit every two years.
- » The Persian buttercup is the first host for the fungus, and it then moves on to the fruit trees.
- » After successfully reproducing in blackthorn (*Prunus spinosa*), the fungus can then go on to other cultivated species.

Survival

- » Mycelium overwinters in the underground parts of plants in the *Anemone* genus (*Persian buttercup*).

Management

- » Raking the orchard's leaves and burning them to ashes.
- » *Prunus spinosa* and other plants have been wiped out, along with any *Anemone* spp.

Bacterial spot, *Xanthomonas arboricola* pv. *pruni*

Symptoms

- » Bacterial spot is a disease that can infect the stems, leaves, and fruits of an infected plant.
- » Small, round, or irregularly shaped lesions that are pale green first emerge on the leaves.
- » The lesions tend to cluster close to the leaf tip when they first appear. In its later phases, the lesion's interior section comes out, leaving the leaf with a jagged or "shot hole" appearance.
- » Miniscule, olive-brown, round dots are the first visible symptoms on the fruit.
- » The patches become darker and lower as the germs grow.
- » The lesions are dispersed around the surface of the fruit, and tiny fissures form at their centres.

Favorable conditions

» Both bacterial reproduction and lesion development require free moistu (dew, rain, irrigation).

» The bacteria are carried from tree to tree by wind-driven rain.

» The sides of the trees that are exposed to the infected winds will suffer mo severe infections.

Survival and spread

» The most common route of introduction and dissemination of this bacter is through infected planting material.

» Gummosis, an ooze produced by bacteria, is spread around by insects, win and rain.

» Additionally, infected harvesting and trimming tools might spread tl bacterium.

Management

» Trees can dry out faster after rain by having their interiors pruned. This als helps keep them from getting sick.

» After the leaves have dropped but before the tree enters dormancy in tl fall, apply one or two foliar sprays of a fungicide containing copper.

» Infection can be kept to a minimum by planting resistant types to bacteri spot. Redskin, Redhaven, Loring, Candor, Biscoe, Dixired, Sunhave Jefferson, Madison, Salem, Contender, Harrow Beauty, and Harrow Diamor are all examples of cultivars that are resistant to the disease.

Peach yellows, *Phytoplasma pruni*

Symptoms

» Newly developing leaves generally have a yellowish tinge on trees affecte by peach yellows, the first sign of the disease.

» A sickle-like look in the immature leaves is also possible.

» Initially, only one or two branches may show symptoms, but as peach yellov spreads, slender, upright shoots (sometimes called witches' brooms) sprou from branches.

» Premature ripening is common, and the resulting bitter taste is unpleasar

Survival and spread

» *Macropsis trimaculata* is the leafhopper responsible for spreading the phytoplasma.

» Grafting or budding contaminated tissues can spread the disease.

» This disease can be transmitted from mother plants to their offspring, the seeds.

Management

» Eliminating infected plants is step one in combating peach yellows.

» Spraying leafhoppers with Neem oil or insecticidal soap once a week until they disappear is the most effective method of control.

» These pests can also be controlled with conventional pesticides like Imidacloprid or Malathion, however their use during bloom will result in the death of honeybees.

» The treatment of peach plants with microinjection capsules of Oxytetracycline has proven effective.

References

Blaauw, B. R., Polk, D., & Nielsen, A. L. (2015). IPM-CPR for peaches: incorporating behaviorally-based methods to manage Halyomorpha halys and key pests in peach. Pest management science, 71(11), 1513-1522.

Davis, M. J., French, W. J., & Schaad, N. W. (1981). Axenic culture of the bacteria associated with phony disease of peach and plum leaf scald. Current Microbiology, 6, 309-314.

Layne, D., & Bassi, D. (Eds.). (2008). The peach: botany, production and uses. Cabi.

Luo, C. X., Schnabel, G., Hu, M., & De Cal, A. (2022). Global distribution and management of peach diseases. Phytopathology Research, 4(1), 1-15.

Margaritopoulos, J. T., Kasprowicz, L., Malloch, G. L., & Fenton, B. (2009). Tracking the global dispersal of a cosmopolitan insect pest, the peach potato aphid. BMC ecology, 9, 1-13.

Pascal, T., Pfeiffer, F., Kervella, J., Lacroze, J. P., Sauge, M. H., & Weber, W. E. (2002). Inheritance of green peach aphid resistance in the peach cultivar 'Rubira'. Plant Breeding, 121(5), 459-461.

Vásquez-Ordóñez, A. A., Löhr, B. L., & Marvaldi, A. E. (2020). Comparative morphology of the larvae of the palm weevils Dynamis borassi (Fabricius) and Rhynchophorus palmarum (Linnaeus)(Curculionidae: Dryophthorinae): Two major pests of peach palms in the Neotropics. Papéis Avulsos de Zoologia, 6(e202060-si.

Wells, J. M., Raju, B. C., & Nyland, G. (1983). Isolation, culture and pathogenicity (the bacterium causing phony disease of peach. Phytopathology, 73(6), 859-86.

Yadav, S., Sengar, N., Singh, A., Singh, A., & Dutta, M. K. (2021). Identification c disease using deep learning and evaluation of bacteriosis in peach leaf. Ecologic; Informatics, 61, 101247.

Zhang, H., Zheng, X., & Yu, T. (2007). Biological control of postharvest diseases c peach with Cryptococcus laurentii. Food control, 18(4), 287-291.

17

Pear

INSECT PESTS

Codling moth, *Cydia (= Carpocapsa) pomonella*

Damage symptoms

- » The nut can also infest apple, peach, and walnut trees in addition to pears.
- » Caterpillars are responsible for the damage they've done.
- » The pinkish-colored larvae begin their lives feeding on the leaves of the host plant before tunnelling into the fruit and feasting on the pulp.
- » Caterpillars produce an inedible amount of frass that they deposit in the tunnels and on the fruit.
- » The infestation causes the fruit to shrivel and fall off the tree early.
- » Between 30 and 70 % of pears lose their marketability due to damage.

Pest identification

- » The eggs, each no bigger than a pinhead, are first transparent and then turn opaque just before hatching.
- » At maturity, the larvae are either pinkish or creamy white in colour, with a brown head, and they can grow to be as long as 1.3 cm.
- » The adult moths of this species are tiny and a murky brown or grey. The forewing tips are covered in chocolate brown spots that give them the appearance of tree bark.

Survival

- » The mature larvae of the codling moth spend the winter encased in thick, silken cocoons in the soil or trash at the tree's base.

Management

- » Pruning effectively helps open up the tree canopy, allowing treatments to more easily reach the larvae inside.
- » Getting rid of the insect reservoirs can be as simple as getting rid of the wild hosts in those abandoned orchards.
- » Use of Entrust with clay (kaolin) in a formulation.
- » Pear trees are treated with Lead arsenate sprays, the first at petal fall (calyx spray) and the subsequent three to four times every 15 days (cover spray).
- » In large-scale orchards, insect populations are typically reduced by disrupting mating behaviour by the release of pheromones across multiple years.
- » *Trichogramma embryophagum* and *Trichogramma cacoeciaepallidum* are examples of parasitoids.
- » Birds are a form of prey (grey tit, *Parus major* and *Passer domesticus*).

Oriental fruit fly, *Dacus zonatus*

Damage symptoms

- » The adult and the larvae are both harmful.
- » The adult insect pierces the fruit, allowing the maggot to crawl inside and feast on the pulp.
- » Fruit that has been attacked by fungus or a pesticide can shrink, develop deformities, and turn a black color.
- » As a result, they end up on the ground
- » Plus, the pear pest can cause problems for other fruit trees besides pears.

Pest identification

- » The adults are a reddish-brown fly measuring 4–5 mm in length with yellowish cross stripes on the abdomen.

Management

- » DDT sprays containing 0.25 % active ingredient are being used to rid the area of adult females that are ovipositing.
- » Before planting, incorporate 400 gm of Dihedron per acre of soil.

» Liquid poison bait is sprayed using a mixture of 21 gm of sugar and 7 gm of 60% Malathion emulsion dissolved in four to five litres of water.

Thrips, *Taeniothrips inconsequens*

Damage symptoms

» Thrips are little insects that feed on plant matter. Their mouthparts can either rasp or sucking.

» Damage to the leaf's underlying cells causes the leaf's surface to wrinkle and take on a silvery grey colour.

» The undersides of the leaves, most noticeably along the mid rib and veins, of moderately affected plants will reveal silvery feeding scars.

» Infested plants have scarred and malformed fruit, stunted young leaves and terminal growth, and a silvery-browning of the leaves overall.

» Developing leaves get twisted in the expanding tips.

» Thrips can also cause fruit damage.

Pest identification

» Nymphs are very lively and their skin is a pale yellow.

» Black or yellow-brown in colour, adults may have white, red, or black patterns. They range in size from 5 to 14 mm and are slender, sucking, rasping insects with fringed wings.

Management

» Spinosad in its Entrust spray formulation.

» The antlion, predatory thrips, coccinellids, anthocorids, lygaeids, and many other insects and animals are among the antlion's many enemies.

Midge, *Contarina pyrivora*

Damage symptoms

» All affected fruits will be slightly bigger than average, rounder than usual, and softer than normal.

» After a few weeks, the developing pear's bottom end (the part farthest from the stem) turns black, and its shape may begin to distort.

» The fruits usually drop to the ground during the months of May and June.

» Maggots of a pale cream colour can be observed moving about inside th immature fruit that has been infected.

» Birds will devour a few of the cocoons.

Management

» Eliminate a food source for maggots over the winter by throwing awa contaminated fruit.

» To prevent maggots from burrowing into the ground from dropped frui spread the polythene as far as the tree canopy reaches.

» Before a flower fully opens, but while the petals are still visible in the tight bu (the white bud stage), spray it with Deltamethrin or Lambda-cyhalothri

» Concorde, Doyenne du Comice, and Onward are less likely to be impacte in comparison to the pear.

» The cocoons are easily destroyed by light cultivation of the soil around pea trees using a trowel or similar tool in the fall or early winter, as this expose the insects to the cold and snow.

DISEASES

Brown rot, *Monilinia* sp.

Symptoms

» As a result of the attack, the leaves droop but do not fall, the blooms darke and dry out, and the branches splay out like a hook.

» The immature fruits wilt, darken, and drop heavily.

» When fruit reaches maturity, the flesh rots and the skin turns a pillow yellowish-grey colour.

» At last, the fruits mummify and stay on the trees, guaranteeing the sprea of the disease into the following season.

Survival

» Spores can survive the winter if they are stuck to a tree through some othe food source, such as a leaf or a mummified fruit.

Management

» Leaves and fruits can be gathered in the fall and burned as fuel.

» Avoid injuring the tree. Protect the injured area of the tree right away with wound-healing paste if you find one.

» Infected fruit should be removed as soon as possible.

» In the fall and winter, spray the Bordeaux mixture two or three times at two week intervals.

» To prevent fruit rot from ever occurring, spray Fenbuconazole-based products on the trees when they are in blossom in the spring.

» Fungus can spread rapidly through close contact between fruits. To facilitate its rapid spread, the fungus causes diseased fruits to cling to healthy ones in the area.

Scab, *Venturia pyrina*

Symptoms

» Weaved dark lesions grow on the leaf blade, twigs, flower pedicels, and cheeks of fruits, and they are superficial, round, and olivaceous.

» It will cause the leaves to turn yellow and fall off early.

» Large, irregular, dull-green, felt-like patches form on fruit when many lesions coalesce.

» The passage of time causes them to age and deteriorate, turning grey and brittle.

» Deformities appear in the fruit.

Survival

» This fungus can spend the winter dormant on old leaves.

Management

» Fallen leaf and fruit collection and incineration.

» As soon as buds have formed, using fermented stinging nettle tea can stop scab from ever developing.

» Bordeaux combination spray has anti-scab properties that can prevent the spread of the disease.

» During three upcoming infection spikes, spray Benomyl or Dodine (delayed dormant stage of flower bud development, pre-pink stage, and calyx stage – after most of the petals have fallen).

> » To prevent scab from overwintering in the garden, collect all of the leave at the end of the season and burn them.

Powdery mildew, *Podosphaera leucotricha*

Symptoms

> » Pear fruits are more susceptible to powdery mildew than the plant's leave or twigs.
> » It leaves behind darkened regions that harden and perhaps split open.
> » The problem is especially severe on pears when they are planted next t orchards of vulnerable apple cultivars.

Management

> » The early detection of disease and the prompt application of fungicide such as Benomyl or wettable sulphur, Myclobutanil (Nova or Rally), o Fenarimol (Rubigan).
> » It is best to start applying fungicides when flower buds are in their tigh cluster stage and keep doing so until growth comes to a complete halt.
> » Tougher varieties include Cvs. Bartlett and Flemish Beauty.

Fabraea Leaf spot, *Fabraea maculate*

Symptoms

> » Leaves, stems, and fruits are all affected by the infection.
> » Spots on leaves manifest as little purple dots before expanding into 6 mn in diameter, deep purple or dark brown infections.
> » Centers of these spots are surrounded by a ring of little black pimple-lik fruiting structures.
> » Black sunken areas emerge on the fruit, causing the skin to split.

Survival and spread

> » Dead, infected leaves are the principal inoculum for the fungus.
> » Rainfall is a key factor in the release and subsequent dispersion of spores.

Management

> » The risk of leaf spot on pears can be drastically reduced by collecting an discarding all falling leaves.

» Ferbam, Ziram, or a Bordeaux mixture applied early has proved effective in preventing and controlling the spread of the disease.

Cercospora leaf spot, *Mycosphaerella pyri*

Symptoms

» It's only in the leaves, so it's not a big deal.

» Spots on mature leaves have a grayish-white centre and a clearly defined edge.

» The lesions, which initially appear as little brown dots on the upper leaf surfaces, eventually grow to be between 1/8 and 1/4 inches in diameter.

» The outer edges turn a deep chocolate brown, and little black bumps form in the middle.

Survival and spread

» Overwintering leaves produce sexual spores that are transferred by winds to emerging leaves.

» New spores are produced in the centres of the grayish-white leaf spots around a month after infection, and these spores are dispersed by rain to other leaves.

» Late summer/early fall is often peak season for these secondary infections.

Management

» Dig up and destroy any leaves or other organic matter that may harbour pests.

» Effective disease control was achieved by spraying 0.1% Carbendazim in the third week of June, 0.3% Zineb 20 days later, and again 0.1% Carbendazim after 20 days (*Shah et al.*, 1985).

Rust, *Gymnosporangium sabinae*

Symptoms

» In Europe, rust is a devastating disease of pear trees when they are grown near Juniper sabianae, the alternate host.

» *Juniper* rust is perennial and looks like apple and quince rust.

» A widespread infection causes the leaves to fall off.

» Early symptoms of rust manifest as yellow lesions on the upper leaf surface.

» Subsequently, lesions could show up on the underside of the leaf.

» The calyx end of the fruit may develop darker blemishes, leading to a distorted appearance.

Survival and spread

» After overwintering on juniper leaves, the spores are released in the spring, typically around May or June, and are carried by the wind to infect pear trees once more.

Management

» Since basidiospores can travel great distances, eradication is uncertain even if infected native cedars (juniper trees) in the immediate area are removed.

» Fumigants are the most effective treatment for rust (Dithane M-45).

Crown and collar rot, *Phytophthora cactorum*

Symptoms

» Cankers, which are decayed patches on the trunk between the ground and the tree's crown, are the telltale signs of collar rot.

» Over time, cankers grow in size and depth, becoming slightly depressed.

» In order to see the tree's bark and roots, the soil around the trunk must be removed.

» The lesions will become coated in a sticky substance during damp conditions.

» Even if these cankers only girdle the tree partially now, they will eventually kill it if the fungus's growth circumstances are maintained.

» The health and vitality of infected trees typically suffers.

Favorable conditions

» Whenever there is an abundance of moisture close to the earth, this sickness will manifest.

Management

» Orchard areas should have light, well-drained soil.

» Phytophthora can be prevented by planting healthy, strong trees.

» In order to keep the graft union above ground, trees should be planted at a shallow depth.

» Dig around the trunks of diseased trees to uncover the cankered wood.

» Irrigations that make sense.

» Ridomil, Subdue, or Aliette, or a fixed copper fungicide (containing 50% metallic copper), can be sprayed on the lower trunk.

Fire blight, *Erwinia amylovora*

Symptoms

» Cankers develop on young shoots and branches.

» At the canker's edge, bacteria congeal into an ooze.

» Ooze attracts insects, which then transport it to the flower's open petals.

» Within 7-10 days of infection, blossoms are ruined.

» The fruit peduncle and then the twig were infected by the bacteria after the blossom was infected.

Favorable conditions

» High humidity, wind, rain, and temperatures between 10 and 30 °C are ideal for the development of this disease.

Survival and spread

» The bacterium can survive the winter in decaying fruit or on fresh stems.

» The insects then transfer the ooze to the flowers they visit, infecting them.

Management

» Sanitize the field by picking up any stray leaves or rotting fruit.

» To get rid of bacterial cankers that formed throughout the winter, prune during dormant months.

» Remove diseased branches by cutting them back 8 to 12 inches.

» Reduce your nitrogen use if possible.

» At 5-day intervals between early and late bloom, use bactericides (Copper fungicides or antibiotics).

» Honeysweet, Kieffer, Orient, Garber, and Douglas are all resistant or tolerant plant cultivars.

Stony pit, virus disease

Symptoms

> » Lots of different viruses are responsible for causing this disease.

> » Prominent lumps can be seen on the surface of the fruit.

> » Pear pit cells merge form a solid, cone-shaped mass in those with rocky p

> » The fruits will get misshapen as it matures.

> » There are a lot of brown and hard spots on the pulp.

> » The fruits lose their edible quality as a result.

> » Grafting can spread this disease.

Survival and spread

> » The virus can be spread through budding, grafting, and root cuttings.

> » There is no evidence that it is transmitted via insects or contaminated seed

Management

> » Only use wood from healthy stock for grafting, roots, or budding.

> » Cut down diseased trees and replace them with pears that have been teste
>   and confirmed to be virus-free.

References

Benbow, J. M., & Sugar, D. (1999). Fruit surface colonization and biological contr of postharvest diseases of pear by preharvest yeast applications. Plant diseas 83(9), 839-844.

Burts, E. C. (1983). Effectiveness of a soft-pesticide program on pear pests. Journ of Economic Entomology, 76(4), 936-941.

Davies, D. L., Guise, C. M., Clark, M. F., & Adams, A. N. (1992). Parry's disease pears is similar to pear decline and is associated with mycoplasma-like organism transmitted by Cacopsylla pyricola. Plant Pathology, 41(2), 195-203.

Horton, D. R. (1999). Monitoring of pear psylla for pest management decisions an research. Integrated Pest Management Reviews, 4(1), 1-20.

Husain, M., Rathore, J. P., Sharma, A., Raja, A., Qadri, I., & Wani, A. W. (2018 Description and management strategies of important pests of pear: A revie J. Entomol. Zool. Stud, 6, 677-683.

Liu, H., Wang, G., Yang, Z., Wang, Y., Zhang, Z., Li, L., ... & Qi, L. (2020). Identification and characterization of a pear chlorotic leaf spot-associated virus, a novel emaravirus associated with a severe disease of pear trees in China. Plant Disease, 104(11), 2786-2798.

Mari, M., Bertolini, P., & Pratella, G. C. (2003). Non-conventional methods for the control of post-harvest pear diseases. Journal of Applied Microbiology, 94(5), 761-766.

Rossi, V., Pattori, E., & Bugiani, R. (2008). Sources and seasonal dynamics of inoculum for brown spot disease of pear. European Journal of Plant Pathology, 121, 147-159.

Shaw, B., Nagy, C., & Fountain, M. T. (2021). Organic control strategies for use in IPM of invertebrate pests in apple and pear orchards. Insects, 12(12), 1106.

Sutton, T. B., Aldwinckle, H. S., Agnello, A. M., & Walgenbach, J. F. (Eds.). (2014). Compendium of apple and pear diseases and pests. The American Phytopathological Society.

18

Pineapple

INSECT PESTS

Mealy bugs, *Dysmicoccus brevipes*

Damage symptoms

- » The nymphs and adults alike feed on the sap from the leaves.
- » Leaf yellowing and wilting are symptoms.
- » Symptoms of mealy bug damage include a reversal in leaf colour from green to red, with the edges reflexing inwards and eventually turning pink.
- » After a long wait, the leaf tips have begun to dry.
- » The badly afflicted plants shrivel up, produce little fruits that never fully mature, and taste terrible.
- » Mealy bugs typically feed on plant roots and leaf axils, but at times of high population, they shift their diet to fruits.
- » The pineapple wilt virus is transmitted by this insect.

Pest identification

- » **Egg:** A typical size for an egg is between 0.3 and 0.4 mm.
- » **Nymph:** They have a thick covering of hairs all over their body.
- » **Adult:** White mealybugs are adults. Soft and convex, with a pinkish colour the adult female body is somewhat rounded. There are also 17 sets of wax filaments that encase their body.

Management

» Grass and other monocot weeds should be eradicated since they are alternate hosts for the insect.

» Ants should be kept under control because they carry mealy bugs from plant to plant.

» Pick healthy suckers and soak them in Methyl parathion or Fenitrothion emulsion (0.05% each) for 15 minutes before to planting.

» Three months after planting, apply 17.5 kg/ha of phorate granules before irrigating the field for rapid translocation of the toxicant.

» Plant hardy crops like Red Spanish and Queen

» Get seeds and other gardening supplies from a nearby, untouched forest.

» Take off the brown leaves at the base.

» Malathion 0.2% should be dipped into the soil around the roots of the plants.

» Dimethoate 2 ml/L or Methyl-demeton 1.5 ml/L sprayed on the area.

» Let out 10 Coccinellid beetles, Cryptolaemus montrouzieri, per tree.

» Anagyrus ananatis and A. kamali are two parasitoids that can be released.

Scales, *Diaspis bromiliae, Melanaspis bromiliae*

Damage symptoms

» Rust-colored patches are an attack symptom.

» It is possible to locate the insect by looking for a protective secretion.

» For feeding, insects use their long, piercing mouthparts to syphon sap from plants.

» If scales find their way onto a pineapple plant in big enough quantities, they can cause significant harm to the foliage, including leaf dieback.

» The scale insect's armour consists of a waxy fluid from the insect and discarded skins. Since the mature bug is hidden deep under the plant's epidermis, controlling the pest requires intervening before the insect reaches this stage.

» During times of high population, scales can accumulate to the point where they completely cover infected plant tissue.

Management

>> Pineapple plants can have their scales eliminated with the help of insecticidal soaps.

>> Little wasps like *Aphytis chrysomphali*, *Encarsia citrinus*, and *E. pernicios* may be able to keep this pest in check.

>> Predators include the thrips and predatory insects *Rhyzobius lophanthae*, *Chilocorus infernalis*, *Pharoscymnus flexibilis*, and *Telsimia nitida*.

hrips, *Holopothrips ananasi, Thrips tabaci, Frankliniella occidentalis*

Damage symptoms

>> Thrips damage plant leaves in order to get sap for their diet. Silvery flecking on the leaf surface is a common sign of deterioration, which can progress to a brown coloration if left unchecked. It's impossible for these leaves to perform photosynthesis properly.

>> An infestation of thrips is indicated by the presence of little black spots on the leaves, which represent the insects' faeces.

>> The majority of thrips will sleep in cracks or along leaf veins. During the day is when you're most likely to see them.

>> They mostly feed on flowers, and their diet contributes to "dead-eye" in the fruit.

>> Crown leaf concentric ring patterns are caused by thrips eating on the crown of fruits.

>> Many plant-borne fungal and viral diseases are spread by thrips.

Management

>> Infestations of thrips can be greatly diminished by removing weeds, rotating crops, and mulching plants.

>> Maybe if we irrigate less, the thrips won't multiply so much.

>> Because they are carried on the wind, thrips populations can be reduced by erecting windbreaks.

>> Garlic and pepper spray can be used to control thrips.

>> The predatory mite, the predatory thrip, *Orius insidiosus*, Coccinellids, etc. are all examples of predators.

Rhinoceros beetle, *Oryctes rhinoceros*

Damage symptoms

» Plant death from adult boring insects near the base of the stem.

Pest identification

» **Grub** - The sluggish, white, 'C'-shaped grub.

» **Adult** - stocky and dark brown or black, and have a horn that rises beyond their heads.

Management

» Burn any pineapple plants that have died.

» Remove the various biostages from the manure pits and dispose of them.

» Adults can be killed by iron catching them at their base on the stem.

» Put up light traps at a density of 1/ha.

» In order to attract beetles, soak 1 kilogramme of castor cake in 5 litres of water and store the pots in a small mud jar.

» *Oryctes* baculovirus and *Metarrhizium anisopliae* are sprayed.

» To catch adults, set up a Rhinolure vane trap.

» Reduviid bug **(***Platymeris laevicollis***)** is a predator that needs to be protected.

» Once every three months apply a 0.1% Carbaryl spray.

» Use a 0.5% Malathion spray.

DISEASES

Basal/fruit rot, *Ceratocystis paradoxa*

Symptoms

» Grey dots on black borders transform into olive brown or white on the leaves

» Symptoms of the disease include tissue dryness and leaf distortion.

» Fruits that start out yellow but ripen into a black colour have water-soaked lesions.

» Those affected undergo a rotting process.

» Invasion occurs when the fungus enters a host organism through a cut or wound and then spreads both between and within its host's cells.

Survival and spread

» The fungus lives as chlamydospores in soil and decomposing pineapp remnants, and plays a crucial role in their decomposition after cultivatio

» Fungi infect plants through open wounds left when cuttings are separate from their parents, eating away at the tender tissue near the stem's base.

» Things that are gathered up in the wet and then piled up are more likely catch some sort of disease while they are waiting for their turn to dry ou Crowns, or tops, used in planting are especially vulnerable.

» Conidia are formed in wet environments and distributed by the wind.

Management

» Suckers for new planting should come from disease-free farms.

» Plant nursery soil is disinfected with either Brassicol or Vapam to elimina the disease-causing organism.

» It has been suggested that, before planting, suckers be doused in a 1 Bordeaux mixture.

» The primary method of disease control is the removal and destruction infected plants.

» Fruits that have been harvested can be protected from stem end rot k leaving them out in the sun for 2 hours.

» Dip the severed end of the fruit stem into a solution of 10% benzoic ac in alcohol.

» Fruits were effectively protected from the disease when they were dipped f 5 minutes in a solution of 1000 ppm Thiobendazole or 2000 ppm Benomy

» During the 4-5 day journey, the stem ends were dusted with 0.1% Benzo acid in kaolin and placed in bamboo baskets to prevent rot. It was determine that the treatment was commercially viable due to its low cost.

» It is recommended to use a 3% Formalin solution to spray the packir containers and baskets.

Heart/stem rot, *Phytophthora nicotianae* var. *parasitica*

Symptoms

» The infection can initiate anywhere on the plant, including the tips, tl base, and the roots.

» This causes the leaves to become dry and wilt, and the veins to turn pink and brown as the process progresses.

» A little pull will dislodge the inner whorl of leaves from the stem.

» Discoloration of the stem's edges between healthy and sick tissue may be brown.

» Poorly drained locations have a higher disease prevalence.

Survival and spread

» The fungal chlamydospore is the principal inoculum, and it can live in the soil or on infected plant detritus for years.

» They can either germinate invisibly, resulting in sporangia, or directly, resulting in hyphae that infect roots and young leaf and stem tissue.

» Soil-dwelling diseases like *Phytophthora* necessitate moist conditions for spore development and infection. Because sporangia formation and the release of motile zoospores require access to free water, infected and diseased soils benefit less from adequate drainage.

Management

» Suckers can be protected from disease by dipping them in a 1% Bordeaux mixture.

» Suckers can be planted in well-drained soil to reduce the spread of disease.

» It is possible to effectively manage the disease with the use of Fosetyl AL dips prior to planting or foliar sprays.

Thielaviopsis rot, *Thielaviopsis paradoxa*

Symptoms

» The sole visible sign is a small darkening of the skin caused by water soaking the epidermis over rotten parts of flesh, which begins at the stem and spreads through most of the flesh.

» The skin over the flesh easily splits with minimal pressure as the flesh softens.

Management

» Don't use any suckers that have come from a contaminated region.

» When planting, utilise raised beds instead of digging holes.

» Enhance the soil's drainage.

» For pre-plant dipping and soil drenching, use either 1% Bordeaux mixtur or 0.25 % Copper oxychloride.

Black spot, *Pencillium funiculosum, Fusarium moniliforme*

Symptoms

» One infected fruitlet (floral cavity) is all it takes to spread the infectio across an entire fruit.

» A brown spot can develop anywhere on the apple, even the centre.

» Browning on the inside of fruit is invisible unless it is sliced into cylindrica shapes.

» It takes about 5-6 days after harvesting before symptoms occur.

Management

» Sprays containing a tank mixture of Benomyl 50% WP and Mancozel 80% WP were applied beginning one week prior to flower induction an continuing for eleven weeks after flower induction, resulting in a significan decrease of 75.8% in the total number of black spots per fruit during storag for fourteen days at room temperature.

Mealy bug wilt, *Pineapple Mealy bug Wilt-Associated Virus* (PMWaV-2)

Symptoms

» In the early stages, you may notice a minor reddening of leaves aroun midway up the plant. The leaves turn pink as they lose their crimson colo become floppy, curl at the edges, and eventually die off at the tip.

» Similarly, the plant's root tissue disintegrates and it begins to wilt. Plant can make a full recovery, albeit they will have smaller, less healthy fruit an smaller, symptom-free leaves.

» In the winter, when development and vitality are suppressed, symptom become more noticeable.

» Disease progression and incidence are both influenced by the age of the plan at the time of the first mealy bug infection, with younger plants showin symptoms two to three months after feeding and older plants taking up t a year longer.

Survival and spread

» The pink mealy bug (*Dysmicoccus brevipes*) is the most likely vector because of its ability to spread viruses that cause the sickness.

» It is likely that infected seeds or potted plants were the source of the outbreak. Once a virus is established, it is spread when mealy bugs feed on tender new growth. Mealy bugs don't fly, yet they can be transported from one plant to another by the ants that follow them around.

» Mealybugs have devoted caretakers in ants. The *Pheidole megacephala*, or coastal brown ant, is active and widespread. Honeydew, secreted by mealy bugs, is a source of nutrition for ants. Mealy bugs are transported by ants, and the insects are defended from predators while in transit. The presence of mealy bugs, the duration of their feeding, and ant activity are all connected to the degree of wilt in a given field.

Management

» Plants must be grown from virus-free material. Through meristem tissue culture, viruses can be eradicated.

» Crowns or slips should be soaked in water heated to 50°C for 30 minutes.

» Crowns and slips may be dipped in a mixture of Diazinon and white oil or Horticultural oil.

» Pruning plants affected with pineapple wilt.

» Close plant spacing and high nitrogen levels reduce the occurrence of pineapple wilt.

» Mealworms can be prevented by a diazinon dip before planting or an aerial spray.

References

Baiswar, P., Borah, T. R. and Singh, A. R. (2021). Postharvest Diseases of Pineapple and Their Management. In Postharvest Handling and Diseases of Horticultural Produce (pp. 287-294). 6000 Broken Sound Parkway NW, Suite 300, Boca Raton, FL 33487-2742: CRC Press.

Chellappan, M., Viswanathan, A., & Mohan, L. K. (2022). Pests and Their Management in Pineapple. Trends in Horticultural Entomology, 689-699.

Kumar, M., Singh, P., Kumar, A., Kumar, S., & Rai, C. P. Management of insect-pests of pomegranate and pineapple. Agriculture Science: Research and Review, 11, 149.

Paull, R. E. and Duarte, O. (2011). Pineapple. In Tropical fruits, Volume 1 (pp. 327-365). Wallingford UK: CABI.

Petty, G. J., Stirling, G. R., & Bartholomew, D. P. (2002). Pests of pineapple. In Tropical fruit pests and pollinators: Biology, economic importance, natural enemies and control (pp. 157-195). Wallingford UK: CABI Publishing.

Rohrbach, K. G. and Schmitt D. (2003). Diseases of pineapple. Diseases of tropical fruit crops, 443-464.

Talib, S. M., Mokhtar, A. S., Wan Abdul Ghani, W. M. H., Jamian, S., Wahab, A. and Aswad, M. (2023). The Insecticidal Potential of Azadirachta indica and Phaleria macrocarpa Plant Extracts against Pineapple Mealybug, Dysmicoccus brevipes (Hemiptera: Pseudococcidae). Egyptian Academic Journal of Biological Sciences. A, Entomology, 16(1), 149-154.

19

Plum

INSECT PESTS

Curculio, *Conotrachelus nenuphar*

Damage symptoms

- » Larvae of the plum curculio feed on the flesh of the fruit.
- » Adult female curculios make feeding and egg-laying holes in fruit by chewing.
- » Infected fruit either falls from the tree or is misshapen (cat-faced) and therefore unsellable if it does not fall.
- » Brown rot fungus can infect fruit at feeding and egg-laying sites.

Pest identification

- » Mature grubs are about half an inch in length, dirty white or yellow in colour, and legless
- » A tiny beetle, measuring only a quarter of an inch in length, the adult has a slender, curled snout.

Survival

- » It spends the winter in the trees and then flies into fruit groves in the spring.
- » Larvae of the plum curculio overwinter in rotting fruit, emerging as adults in the spring to infest new crops.

Management

- » Frequently pick up and dispose of rotting fruit.

» Shuck-split is the time to apply an effective insecticide; subsequer applications should be made 10–14 days apart, and the final applicatic should be made about 30 days before harvest.

» To prevent the adults from feeding and laying eggs in the fruit, the: treatments should begin when 90% of the petals have fallen.

Oriental fruit moth, *Cydia funebrana*

Damage symptoms

» The first generation of larvae attacks the sprouts and feeds inside, wreakir havoc on the newly planted orchards.

» The third and fourth generation larvae are the most destructive to the fruit

» By chewing irregular galleries around the stone, larvae gain access to tl fruit through the peduncle.

» The fruits are especially vulnerable to moth damage, but the sprouts an leaves are affected as well.

» Fruit that is attacked stops developing and eventually falls to the ground

» The larvae will keep on eating the rotting fruit.

Survival

» The larvae overwinter in a silken cocoon in the crevices of the bark and tl species produces three to four generations yearly.

Management

» The infected plants and fruit were cut down and destroyed.

» The setting of traps that draw in prey.

» Sprays containing Malathion, Permethrin, or Lambda cyhalothrin shoul be used.

» Therapeutic interventions including 20 MG Mospilan, 50 CS Karate Zeo: 50 CS Decis Mega, 480 SC Calypso, 240 SC Affirm, 10 CE Laser, an SC 240 SC Faster.

Tree borer, *Synanthedon exitiosa*

Damage symptoms

» The peach tree borer larva typically feeds from the tree's main roots to withi about 10 inches of the tree's trunk.

» Gum in large quantities, often combined with frass (the sawdust-like excrement of insects).

» Just a few larvae can kill a young tree.

» An increased number of larvae is tolerable only on older trees.

Pest identification

» The adults are clear-winged moths, with the female being metallic blue-black save for a reddish-orange stripe across her abdomen. Males are distinguished by their black plumage, which is adorned with small yellow stripes on the abdomen and longer yellow stripes along the back, towards the wing's origin.

» The full-grown length of the larva (immature stage) is between 1 and 114 inches. The creature's body is a milky white, while its head is a chocolate brown.

Survival

» It overwinters as a larva.

Management

» Mist the ground around tree bases with Permethrin.

Scale insect, *Sphaerolecanium prunastri*

Damage symptoms

» This species feeds on many different kinds of plant life, including fruit trees, vines, and so on.

» Adults and larvae spread necrosis by colonising tree limbs and plant foliage.

» In addition, they can cause leaf deformations and even drop if they assault the foliage.

» Their sugary droppings smother the affected plants, which in turn encourages the growth of certain phytopathogenic fungi.

Survival

» It produces only one generation every year, and its larvae spend the winter attached to the bark of the trees it attacks.

Management

» Scrubbing the tree limbs clean.

» Care plans that include administering Mospilan 20 SG, Nuprid AL 20 SC, Decis Mega EW 50, Calypso 480 SC, and Faster 10 EC.

Leaf curl aphid, *Brachycaudus helichrysi*

Damage symptoms

» It seeks out and destroys the newly sprouted foliage.

» As a result, it drains the vitality from developing organs and tissues.

» Curled leaves are a sign of an infestation of this bug.

» Early in the year, infestations typically affect only one or a small number of limbs.

» This aphid secretes a lot of honeydew.

» Aphid populations can stunt tree development and lower fruit sweetness

» When pest infestations become severe, they stunt plant growth and prevent flowers from opening.

Management

» To expedite leaf drop, use zinc sulphate (36%), 20-50 kg/ha, in early to mid-October. Without the tree's leaves, the aphids can't reproduce.

» Late fall/early dormancy (November 1) chemical treatment with Phosmet (Imidan) 70W, Diazinon 50WP, or Fenvelarate is an effective method of pest management.

» Apply Dimethoate (Rogor 30 EC) @ 2 ml/L or Monocrotophos (Nuvacuron) @ 1.5 ml/L to the plants during bud burst and again at fruit set.

» The parasitoid *Aphidius colemani* is mostly responsible for the high rates of parasitism seen in this species of aphid.

» The lady beetle, green lacewing, brown lacewing, syrphid fly, and soldier beetle are all significant predators.

Spider mites, *Panonychus ulmi, Tetranychus urticae*

Damage symptoms

» Indirectly harming fruit, mites create stippling, bronzing, and even leaf drop as a result of eating on leaves.

» Because photosynthesis is slowed down, vitality and production are diminished.

Pest identification

- » Miniscule (0.5 mm or less) adults. The color might range from a pale greenish yellow to a deep red or a rich brown.

Survival and spread

- » Under bark, on smaller branches or fruit spurs, or in the ground cover are ideal wintering spots for mites and their eggs or adult females.

Management

- » Mite populations could be reduced through the use of cultivation or grasses to suppress weed growth.
- » A vigorous stream of water may be able to remove the mites from the tree.
- » Trees are particularly vulnerable when drought-stressed, so make sure you water them regularly.
- » Too much nitrogen might actually attract mites, so try to limit your treatments.
- » Exclusively use sulphur compounds and horticultural mineral oil (for European red mite only) when plants are dormant or nearing dormancy.
- » During the growing season, use insecticides such as wettable sulphur, azadirachtin (Neem oil), gamma-cyhalothrin, insecticidal soap, pyrethrins, spinosad, and resmethrin.
- » Predatory phytoseiids virtually usually keep mite populations in check.

DISEASES

Bladder plum gall, *Taphrina pruni*

Symptoms

- » The unripe fruits turn a yellowish green and are larger than the mature ones.
- » The attack has distorted the fruit, and the stone's protective rind has been shattered.
- » The fruiting bodies of the fungus, represented by the grey fluff, occur on the surface of the fruits.
- » After around four weeks, the fruits wilt and drop dramatically.

Favorable conditions

» The moist and cold springs are ideal for the development of the sickness.

Survival and spread

» The fungus survives the winter as mycelium attached to tree bark.

Management

» Use of tolerant strains;

» Apply 4 applications of Bravo 500 SC, Score 250 EC, Polyram DF, Merpan 50 WP, and Systhane Plus 24 E if the weather permits.

Shot hole disease, *Stigmina carpophila*

Symptoms

» Brown spots, in the shape of a circle, emerge on the leaves and eventually separate from the remainder of the leaf.

» Dot-like structures, framed in crimson and lilac, can be seen on the fruits. As a result, the pulp of the fruits loses its chewiness and flavour.

» The brunifications on the sprouts are caused by the fungus, and the wounds it creates eventually bleed glue.

» This is a particularly perilous sort of attack since it kills off the young plants just as they start to bear fruit.

Survival and spread

» As a mycelium on the bark of the branches, the fungus can survive the winter.

» The fungus's spores guarantee its continued spread, and the glue that seeps from open wounds may withstand a harsh winter.

Management

» Eliminating and then burning the diseased branches.

» Mastic glue will be applied to the cuts that were made during pruning.

» After the trees have been pruned, a 4-5% solution of Bouillie Bordelaise will be administered as a treatment.

» Sore 250 EC, Folpan WDG, and Rovral 500 SC are applied as a treatment during the vegetative growth phase.

Brown rot, *Monilinia laxa*

Symptoms

- » Significant losses occur during cold and wet years since this disease affects all of the tree's aerial organs.
- » Young flowers and their supporting stems and leaves perish.
- » Cankers on twigs are tiny and tan, with dark margins.
- » Flower bases tend to exude a sticky substance.
- » Flowers with brown spore masses, a sign of high humidity.
- » Brown rot spreads quickly on infected fruit, but small necrotic spots can also appear.
- » Finally, the diseased fruit mummifies and stays on the tree, where it can spread to other trees the following season.

Favorable conditions

- » Wet spells encourage blossom and twig blights.

Survival and spread

- » Fungal spores can be found in dead fruit, ruined flowers, cankers, and infected branches.

Management

- » Taking preserved berries and apples off the tree.
- » To remove diseased branches.
- » Assuring sufficient water and fertiliser levels to alleviate plant stress.
- » It's best to forego sprinkler irrigation in order to keep the foliage and blossoms dry and disease-free.
- » Use of the necessary safety Timing the application of copper fungicides so that they are sprayed when the vulnerable floral portions are exposed or after a period of rainy weather is optimal.

Powdery mildew, *Podosphaera tridactyla*

Symptoms

- » Attack symptoms manifest as white patches on the foliage of newly emerged plants.

» These grow to cover the entire leaf.

» In later stages of the disease, the mycelium looks dusty and grey.

» Attacked tissue dries out and develops wrinkles.

» The young fruits are also susceptible to the fungus. White mycelium felt protects these.

» The fruit spoils and breaks apart.

Management

» Implementation of disease-resistant plant strains.

» The intruders' sprouts and fruit will be burned.

» Medications such as Systhane Plus 24 E, Thiovit Jet 80 WG, Kumulus DF Topas 100 EC, and Karathane M 35 CE.

Black knot, *Apiosporina morbosa*

Symptoms

» Protrusions (knots) on a tree's wood that can be up to 30 cm (12 inches in length

» Initially olive green and corky, knots eventually darken to a brittle black.

» The length of knots increases annually.

Favorable conditions

» In the second year, infections on the new growth following rainstorms and the quick development of knots are major problems.

Management

» Remove branches from the orchard that have knots 8-10 cm (3-4 in) below the swelling.

» Knots should be cut out of older branches, along with two cm (about an inch) of tissue on either side of the knot. Midsummer is the best time of year to get rid of knots.

» If effective fungicides are available, they can be applied during the shoot elongation stage to prevent disease.

Plum pockets, *Taphrina* **sp.**

Symptoms

> » This fungus can cause symptoms on leaves, stems, and fruits, but the latter are where you'll see them most clearly.

> » Six to eight weeks after bud break, the symptoms appear on all plant parts.

> » The fruit gets bloated, wrinkled, and deformed, sometimes growing to be 10 times its original size.

> » All fruits have soft, hollow centres that may or may not be home to a pit or stone.

> » The fruits are variously referred to as "bladder plums," "fake plums," and, more commonly, "plum pockets" when they are still green.

> » The most typical symptoms of leaf and fruit diseases are twisting and curling, however these may not always be present.

Management

> » New trees should be of resistant varieties.

> » When applied in late fall or early spring, just one spray of fungicide is all that's needed to keep problems at bay until the buds open in the spring.

> » Effective treatments include the Bordeaux combination, chlorothalonil, and liquid Lime sulphur.

Crown gall, *Agrobacterium radiobacter* pv. *tumefaciens*

Symptoms

> » Root and stem tumours are common.

> » Small and pliable at initially, over time they harden into something resembling wood.

> » There is a wide range in tumour size and shape.

> » Tumors produce structures resembling leaves, buds, and sprouts.

> » The tumour cells themselves are oversized and misshapen.

> » The tree's defences against the microorganisms are breached by the worms, hail, etc.

Favorable conditions

- » Temperatures between 22 and 30°C, with a humidity of 80%, are ideal for the development of this disease.
- » Crown gall is a disease that first infects a plant at the location of a wound.
- » Soils that drain poorly, are high in alkalinity, and have been damaged by nematodes are ideal conditions for the spread of disease.

Management

- » Using only tested and verified disease-free plant material in your plantings.
- » Pruning and planting plums on soil that drains well.
- » Planting plums in a non-host rotation after treating diseased crops.
- » Decontaminating the shears and tools before moving them along from one tree to another.
- » Treatment involves removing diseased branches to a healthy region, using Bouillie Bordelaise 4-5%, and dressing wounds with resin.
- » Roots of the seedlings will be submerged in a solution of either 1% Bouillie Bordelaise or 0.025% Topas 100 EC.
- » Cucumber-based treatments administered during the growing season.

Bacterial canker, *Pseudomonas syringae* pv. *morsprunorum*

Symptoms

- » Little wet stains form in a circular pattern on the leaves.
- » In high humidity, a white to yellow, viscous pellicle forms on the spots which is bacterial exudate.
- » The damaged tissue on the leaf falls off once it has dried. This will result in a perforated appearance on the leaves.
- » Defoliation of trees is a function of how severe an attack is.
- » Small, reddish-brown dots are seen on the fruits. Because of the assault fruit become misshapen and the pulp can shatter.
- » Long patches develop on the sprouts, the bark darkens and eventually dies.
- » Infected wounds often produce a thick, sticky fluid caused by bacteria.
- » Wounds tend to worsen every year and may eventually develop into full blown malignancy.

» The bacteria enter the tree through the stomata, although grafting is where the most majority of diseasees originate.

Management

» Use of viable grafts in reconstruction.

» The use of Bouillie Bordelaise WDG, Copernico Hi-Bio, Melody Compact 49 WG, Funguran OH 50 WP, and Champ 77 WG for various therapies.

Bacterial spot, *Xanthomonas campestris* pv. *pruni*

Symptoms

» Leaves' undersides may develop water-soaked, angular, grey lesions that, if the lesion centre drops out, give the appearance of shot holes.

» Large numbers of lesions on leaves can cause the leaves to get chlorotic and fall off the tree.

» Cankers appear on twigs as dark areas around buds that never bloom or as raised blisters.

» Complete losses of the fruit crop may occur in years of extreme disease.

» Small brown spots that leak water and sometimes gum appear first on fruit.

Favorable conditions

» Rainfall that occurs often during the late bloom and early petal drop stages raises the risk of infection in the fruit and leaves.

» When the weather is hot and dry, infections are uncommon.

Management

» Do not plant disease-prone cultivars in regions where they have been documented.

» In order to avoid the difficulty of eradicating a disease once it has become obvious, it is recommended to apply a protective copper spray either in the fall before the leaves fall off or in the spring before the plants begin to actively grow.

» Peach trees have a high sensitivity to Copper, so use caution around them.

Plum pox, *Plum pox virus* (PPV)

Symptoms

- » In the beginning of summer, the leaves develop circular discolorations that could later turn necrotic.
- » The unripe fruit is marked with yellow, round dots.
- » Fruit that has been assaulted will ripen early and fall from the tree.
- » The stone's circular spots appear if the attack is really harsh.
- » Chlorotic dots, rings, and lines show on leaves as a pale green colour.

Survival and spread

- » Aphids are the vector for the virus.
- » Viruses can be transmitted from plant to plant by aphids, cicadas, and pollen.

Management

- » Put in only approved, healthy plant material.
- » Space separating the many stone fruit orchards.
- » The elimination of unwanted vegetation around the orchard, specifically weeds and shrubs.
- » Take out the diseased plants in the orchard.
- » Use of chemical sprays for aphid control may actually increase the time it takes for the virus to propagate.
- » C5 is a genetically modified plum cultivar that is immune to the plum pox virus.

References

Birwal, P., Deshmukh, G., Saurabh, S. P., & Pragati, S. (2017). Plums: a brief introduction. Journal of Food, Nutrition and Population Health, 1(1), 1-5.

Cambra, M., Capote, N., Myrta, A., & Llácer, G. (2006). Plum pox virus and the estimated costs associated with sharka disease. EPPO bulletin, 36(2), 202-204. Jaastad, G., RÃ¸en, D., Bjotveit, E., & Mogan, S. (2004). Pest management in organic plum production in Norway. In VIII International Symposium on Plum and Prune Genetics, Breeding and Pomology 734 (pp. 193-199).

Kutinkova, H., & Andreev, R. (2008). Possibility of reducing chemical treatments aimed at control of plum insect pests. In IX International Symposium on Plum and Prune Genetics, Breeding and Pomology 874 (pp. 215-220).

Norton, M. (2007). Growing prunes (dried plums) in California: an overview.

Okie, W. R., & Hancock, J. F. (2008). Plums. In Temperate fruit crop breeding: Germplasm to genomics (pp. 337-358). Dordrecht: Springer Netherlands.

Okie, W. R., & Hancock, J. F. (2008). Plums. In Temperate fruit crop breeding: Germplasm to genomics (pp. 337-358). Dordrecht: Springer Netherlands.

Posnette, A. F., & Ellenberger, C. E. (1957). The line-pattern virus disease of plums. Annals of Applied Biology, 45(1), 74-80.

Sano, T., Hataya, T., Terai, Y., & Shikata, E. (1989). Hop stunt viroid strains from dapple fruit disease of plum and peach in Japan. Journal of general Virology, 70(6), 1311-1319.

Vincent, C., Chouinard, G., & Hill, S. B. (1999). Progress in plum curculio management: a review. Agriculture, ecosystems & environment, 73(2), 167-175.

20 Pomegranate

INSECT PESTS

Anar butterfly, *Virachola isocrates*

Damage symptoms

- » Pomegranate larvae can eat their way through up to half of a crop's fruit.
- » Female butterflies lay their eggs singly on the calyx of flowers or the ski of tiny fruits.
- » Caterpillars, once they hatch, make holes in ripening fruit to eat the pul and seeds inside.
- » Because of this, fruit-rotting microbes and fungus will be able to enter th fruit and do their damage.
- » Signs of damage include a foul odour and caterpillar droppings at the entr points.
- » When this happens, the fruit falls to the ground.

Pest identification

- » **Eggs** - Individual eggs are placed on young vegetation such as leaves, stalk and flower buds.
- » **Larvae** - An adult larva is around 16-20 mm long and has a short, robust body covere in short hair and white patches all over its body.
- » **Adult** - Females are more brownish violet than males, although both sexes ar glossy blue as adults, with a prominent orange V-shaped patch on the forewing

Management

» Get rid of and throw away any contaminated produce (fruits with exit holes).

» No need for pesticides or herbicides because weeds can replace unwanted hosts in a garden.

» When the fruit is about 5 cm in diameter, cover it with polythene bags.

» Keeping tabs on adult behaviour is easy with a light trap set up at a rate of 1 per hectare.

» At the flowering stage, spray NSKE 5%, Neem oil 3%, or Neem formulations 2 ml/1.

» When more over half of the fruits have set, spray them with Decamethrin 2.8 EC at a rate of 1 milligramme per litre.

» Apply Carbaryl 50 WP at 4g/L, Fenvalerate 20 EC at 0.25ml/L, or Quinalphos 25 EC at 2g/L every two weeks throughout the dry months.

» Apply Dimethoate at a rate of 1.5 ml per litre, or ETL, of 5 eggs per plant.

» Eliminate any weeds that have flowers, especially any members of the Compositae family.

» Spread *Trichogramma chilonis* at a rate of 2.5 million per hectare four times, with 10-day intervals between each release.

» *Bacillus thuringiensis* sprayed at a concentration of 1 g/L once weekly has been shown to be effective.

Shot hole borer, *Xyleborus perforans*

Damage symptoms

» Adult beetles cause damage by boring holes in the plant's roots and, eventually, its lower trunk.

» Nesting occurs in the holes.

» Mild yellowing of a lateral branch on one or more trees, typically in a contiguous area, is the first sign of infection in an orchard.

» After only a week, the entire tree has begun to turn yellow and the branches have dried out.

» As a result of the infestation, the trees may produce an abnormally large number of small, unripe fruits.

» Small pinholes can be seen in the main trunk a foot above the ground.

» The tree's xylem and phloem were severed by the perforations.

» Within a month's time, the adults will have migrated to the nearest thriving trees.

» The entire orchard might be contaminated in as little as three to six month if precautions are not taken now.

Pest identification

» Oval or spherical, shining, and iridescent white, shot hole borer eggs are a distinctive feature of this insect.

» White, legless, and up to 4 mm in length, the larvae are a common insec food source.

» About 2-3 mm in length, the adult is cigar-shaped and can range in colou from black to reddish brown. Their head capsule is shortened, and they lack a cranium.

Management

» Keep the soil scraped and aerated to prevent waterlogging.

» Soak the ground surrounding the tree's main trunk with a solution o Chlorpyrifos (2.5 ml) and Tridemorph (1.0 ml/L). We recommend utilising 2-3 litres of mixture per tree.

» After three weeks, spray again with Dimethoate 30 EC @ 2 ml/L + Carbendazim 1g/L (2- 3 L of mixture/tree with either of the above fungicides).

» If the insect problem persists after a month, repeat the drenching procedure

» Use 3 ml/L of Azadirachtin (0.15%) around the main trunk if the infestation level is low.

» All parts of the infected tree, including the roots, should be burned.

» Chlorpyrifos 2.5 ml/L should be poured into the holes left by uprooted trees

Stem boring beetle, *Coelosterna spinator*

Crop losses

» Vine yield decreased by a mean of 5.05 kg/vine and a total of 3475.75 kg/ acre due to borer damage.

Damage symptoms

» After two weeks, the eggs hatch into grubs that eat their way through the tree's main stem, main branches, and minor branches.

» Symptoms of a diseased stem include yellowing, drying, and eventual dieback of its leaves.

» When trees are infected with diseases twice, it causes their deaths to progress slowly.

Pest identification

» Adult beetles are 30–35 mm in length, with a pale yellowish-brown body and light grey elytra.

Management

» Check the orchards on a regular basis to look for dried branches.

» Inserting a hooked wire down the hole will suffocate the stem borer grubs.

» Set up the light traps with a 200-watt sodium lamp to lure in and kill the adults.

» After filling any holes or excreta/gummosis with clay, inject 5-10 ml of Dichlorvos 76 EC @ 4 ml/L into the branches using a disposable syringe (needle-free).

» Remove the dead wood from the branch and swab the exposed cut end with Copper-oxy chloride 50% WP.

» The nearby trees can be treated with Quinalphos 25 EC @ 2 ml/L or Chlorpyrifos 20 EC @ 2.5 ml/L.

» A preventative swab on the main trunk using the IIHR swab mixture (discussed under grapes) should be applied in May or June.

Thrips, *Rhipiphorothrips cruentatus, Scirtothrips dorsalis*

Damage symptoms

» Curling of the leaf tips and flower drop are both symptoms of an infestation by these piercing and sucking insects.

» They create scab and lower the price of fragile fruits when they feast on them.

» Both nymphs and adult's lacerate buds, flowers, leaves, and fruits to extract their juices.

» Thrips feed on the underside of leaves, rasping the surface and sucking the leaking cell-sap, causing the leaf tips to turn brown and curl, which in turn causes the flowers to dry up and fall off.

» Scab forms when fruit is scraped, decreasing its value.

Pest identification

» **Egg:** Eggs are soiled white beans in shape.

» **Nymphs:** Nymphs are reddish when they first hatch, and they mature into a yellowish brown colour.

» **Adults:** *Rhiphiphorothrips cruentatus* adults are 1.4 mm in length and have a lengthened, slender body that is blackish brown with yellowish wings. Straw-yellow *Scirtothrips dorsalis*

Favorable conditions

» From July through October, with the peak occurring in September, this pest is most prevalent.

Management

» Chilies and onions should not be planted close to one another.

» Remove and destroy affected plant parts.

» Use of blue sticky traps @ 1 trap / 10 plants.

» Keep the trunk basins clean.

» Spray with Acetamiprid 20 SP @ 0.005% to 0.01% i.e. 25 to 50 g /100 L or Spinosad 45 SC @ 0.25 ml/L i.e. 25 ml/100 L or NSKE 5% or Verticillium lecanii (2x108 cfu/g) @ 200 g / 100 L starting from prior to flowering at the interval of 10 days.

» It's recommended to spray Chlorpyriphos at 0.02%, Imidacloprid at 0.04%, Deltamethrin at 0.15%, and Dichlorovos at 0.05%, either as a preventative measure or after noticing symptoms.

» Before the flowers open, give them a 0.06% spraying of dimethoate. Spraying Methyl oxydemeton 0.05% twice, once before and once after fruit set, is recommended under extreme conditions.

» Apply 0.5% Multineem spray if needed.

Aphid, *Aphis punicae*

Damage symptoms

» Adults and nymphs alike cause shoot shrivelling and fruit drop by sucking sap from leaves and stems.

» Deterioration of leaf coloration.

» The withering of the plant's last shoots.

» When they are particularly severe, honeydew and sooty mould tend to accumulate.

Pest identification

» Aphids can be either winged or unwinged and are a pale green colour.

Management

» Gather up the broken pieces of the plant and throw them away.

» Set up sticky yellow traps.

» Spraying 2 ml/L of Dimethoate 25 EC or Methyl demeton 30 EC on the emerging shoots.

» We find that NSKE 5% effectively reduces aphids without harming their natural predators.

» There are many predators of aphids, including the coccinellids *Scymnus castaneus, S. sexmaculata, S. latemaculatus, and C. sexmaculatus,* and the syrphid Paragus serratus.

» When the flowers of *Chrysoperla carnea* first begin to open in April, you should release 15 first-stage larvae per flowering branch over the course of four separate releases spaced 10 days apart.

Tailed mealy bug, *Ferrisia virgata*

Damage symptoms

» Causes damage to plant life and fruit production

» The premature ripening and falling off of fruits.

Pest identification

» **Nymph:** Nymphs tend to be a creamy yellow or a very light white.

» **Adult:** Female adults are apterous and have a long, slender body that is covered in white waxy secretions and has a pair of wax filaments at the tai

Management

» The infected plant components must be gathered and disposed of.

» Remove and discard diseased plant components.

» Take out the backup hosts.

» Using a spray bottle, mix 2 ml of Triazhophos with 5 ml of Neem oil pe litre of water, and 1.5 ml of Phosalone 35 EC with 5 ml of Neem oil pe litre of water.

» Use 2 ml/L of monocrotophos 36 WSC or methyl demeton 25 EC as a spra

» Spraying 1 ml of Dichlorvas 76 WSC and 25 gm of fish oil rosin soap pe litre

» After removing infected branches, spray the area with a mixture of Chlorpyrifo 20 EC at 2.5 m/L and Dichlorvos 76 EC at 0.75 ml/L.

» Release 10 *Cryptolaemus montrouzieri* beetles per infested tree if the infestatio is localised.

» Spread 2 g/L of *Verticillium lecanii*.

Fruit borer, *Dichocrocis (Conogethes) punctiferalis*

Damage symptoms

» When young, fruits are attacked by a caterpillar.

» Consumed for its internal organs as food (pulp and seeds).

» The fruits wither and drop off before they ripen.

Pest identification

» Larva: light green with a pinkish undertone, fine hairs, dark head and prothoracic shiel

» Adult: Moth with black dots on its wings and body

Management

» Recover the spoiled fruit and throw it away.

» Weed plants are used as a substitute in organic farming.

» Check on the goings-on of grownups by setting up a light trap at a rate c 1 per hectare.

» Malathion 50 EC 0.1% or Dimethoate 30 EC 0.06%, sprayed twice during the stages of flowering and fruit setting.

Whitefly, *Siphoninus phillyreae*

Damage symptoms

» Adults and juveniles alike feed on leaf sap.

» Honey dew is produced by them, which encourages the growth of sooty mould.

» Deterioration of leaf coloration.

» Leaves turn brown and fall off.

Pest identification

» **Nymph** - wafer-thin glass rods on the sides of the body.

» **Adult** - White and adult, they seem to be most active in the morning.

Management

» Cleanup on the field.

» Elimination of parasitic plants.

» Set up of yellow sticky traps.

» Use NSKE 5% or 3% Neem oil as a spray.

» The Coccinellid predator, *Cryptolaemus montrouzieri,* and the lace wing fly, *Mallada astur,* were released as predators.

» Putting the parasitoids *Encarsia haitiensis* and *E. guadeloupee* out into the wild.

DISEASES

Leaf spot, *Cercospora punicae*

Symptoms

» The foliage and fruit have zoned spots of light brown.

» Twigs develop black, elliptical patches.

» When this happens, the damaged parts of the twigs become depressed and flattened, with a raised edge.

» Such branches eventually wither and die.

» Total plant death can occur in extreme instances.

Favorable conditions

» Heavy precipitation and soggy soil are both conducive to the spread of disease.

» From September through November, the sickness is at its peak.

Survival and spread

» Conidia are the infectious form and are carried by the wind to new hosts.

Management

» Diseased branches can be removed and destroyed through pruning and destruction to help lower the inoculum.

» Infected fruit needs to be thrown away immediately.

» Thiophanate methyl 0.1%, Chlorothalanil 0.2%, or Mancozeb 0.25 % spray is suggested.

Anthracnose, *Colletotrichum gloeosporoides*

Symptoms

» Small, dull violet or black patches with yellow edges are a telltale sign of this disease.

» Later on, the spots start to merge together and grow in size.

» Dying and falling off, infected leaves turn a sickly yellow.

» It's not just bread that fungus may infect; fruit can get it too.

Favorable conditions

» August and September, with their high humidity and 20°C to 27°C temperatures, are peak months for the disease.

Survival and spread

» Discarded plant matter is the primary vector for the spread of plant diseases.

» Conidia distributed by the wind as a secondary vector.

Management

» We need to provide more room between trees and cut them down every year.

» The infected branches and leaves must be thrown away properly.

» Once every two weeks, spray either 0.1% Carbendazim + 0.2% Copper oxychloride, 0.1% Thiophanate methyl, 0.2% Mancozeb, or 0.15% Kitazin.

Wilt, *Ceratocystis fimbriata*

Symptoms

» Leaves on one or more branches began to yellow and wilt, and then the entire shrub died suddenly three to four weeks later

» When the bark was peeled away from sick plants, the roots underneath revealed a brownish-black colour and showed signs of irregularly shaped lesions.

» Some of the leaves on the affected plant's twigs or branches may become yellow, and those leaves will eventually droop and dry out. In a matter of months to a year, the entire tree will have perished.

» Dark grayish-brown staining of the wood is visible when a cross-section or longitudinal slice is taken from an affected tree.

Favorable conditions

» Wetter, heavier soil is more conducive to disease.

» Constant drizzle with temperatures between 18 and 30 °C.

Survival and spread

» Primary source of inoculum: Inoculum is typically obtained from chlamydospores in the soil.

» Secondary source of inoculum: Conidia and water can also serve as a secondary vector.

» Infected seedlings are a vector for the pathogen's spread.

Management

» Sanitation procedures, such as cutting down infected trees and burning them, and treating the soil with formalin (20 ml/L), can halt the spread of the disease.

» The effectiveness of Propiconazole 0.2% was higher than that of Difenoconazole 0.2%.

» The bioagents Trichoderma harzianum and Pseudomonas fluorescens at a 1×10^6 multiplicity of infection were similarly efficient.

Fruit rot, *Phytophthora palmivora*

Symptoms

- » The only time this fungal disease becomes an issue is in poorly drained soils or during periods of severe rain.
- » The fungus causes decay and death in plant tissues, specifically in the bark and roots. Water-soaked light-brown lesions that spread inward from the leaf's perimeter and eventually turn dark brown and watery.
- » The plant as a whole will perish as its leaves droop, turn yellow, and fall off.
- » In humid, damp environments, such as those found in many nurseries, this can also become an issue.
- » Initial symptoms appeared as water-soaked, spherical, light-brown lesions in the centre or bottom of fruits.
- » These lesions quickly grew larger on the surface and deeper within the fruit, resulting in decay throughout.
- » Dense orchards and fruits that were in close proximity to the ground were found to have the highest prevalence.
- » However, some dark brown, mummified fruits managed to cling to the trees after the rest of the rotten fruit had fallen.
- » The mycelium, which grows white and fluffy, is best visible on the tips of the late-blooming flowers. Flowers like that eventually withered and fell to the ground.

Favorable conditions

- » Disease prevalence was higher in orchards with dense canopy and proximity to heavy rain.

Management

- » Reduce watering and enhance soil drainage.
- » For the most reliable drying results, use drip irrigation and water first thing in the morning.
- » Recover diseased produce and throw it away.
- » It is recommended to use Copper oxychloride or Dithane M-45 to treat infected plants after they have been extensively pruned.

» Chlorothalonil, Copper, and Mancozeb are all effective protectors; Metalaxyl and phosphoric acid are alternatives (systemics).

Bacterial blight, *Xanthomonas axonopodis* pv. *punicae*

Symptoms

» Various plant parts, including leaves, stems, branches, and fruits, are susceptible to infection by bacteria.

» The disease manifests as brown to black spots around the nodes of the stem, which eventually girdle and crack, leading to the collapse of the branches.

» The pericarp develops brown to black patches with L- or Y-shaped fissures.

» Fruit spots are elevated and have an oily appearance.

» Fruits may break extensively in extreme circumstances.

Favorable conditions

» Conditions of high heat and low humidity aid in the spread of the disease.

» Increases in daytime temperature (38.6°C) and relative humidity (30.4%) in the afternoon, together with gloomy skies and sporadic rain, helped spark and spread the disease.

Survival and spread

» The tree serves as a host for the bacterium. In the off-season, it can live for up to 120 days on dead leaves.

» Infected tissue is typically spread by the use of cuts. Disease is dispersed via rainfall blown around by the wind.

Management

» Plant only disease-free seedlings.

» Use proper field sanitation measures (such as collecting and burning damaged leaves, stems, and fruits) to stop the spread of disease.

» Sprays of Bromopal (0.5 g/L) or Copper oxychloride (2 g/L) should be administered every 10 days initially.

» Diseased leaves should be sprayed with a 1% Bordeaux mixture prior to being pruned.

» Kill microorganisms on fallen leaves by dusting tree basins with 20-25 kg/ha of bleaching powder.

» Paste 0.5 g/L Bromopal (Bacterinashak or Bitrenetol) or 3 g/L Coppe oxychloride in red sandy loam soil on sick stems after trimming.

References

Ananda, N., Kotikal, Y. K., & Balikai, R. A. (2009). Management practices for majc sucking pests of pomegranate. Karnataka Journal of Agricultural Science 22(4), 790-795.

Balikai, R. A., Kotikal, Y. K., & Prasanna, P. M. (2011). Status of pomegranate pest and their management strategies in India. Acta horticulturae, 890, 569-583.

Cocuzza, G. E. M., Mazzeo, G., Russo, A., Giudice, V. L., & Bella, S. (2016 Pomegranate arthropod pests and their management in the Mediterranea area. Phytoparasitica, 44, 393-409.

Deshpande, T., Sengupta, S., & Raghuvanshi, K. S. (2014). Grading & identificatio of disease in pomegranate leaf and fruit. International Journal of Compute Science and Information Technologies, 5(3), 4638-4645.

Faria, A., & Calhau, C. (2011). The bioactivity of pomegranate: impact on healt and disease. Critical reviews in food science and nutrition, 51(7), 626-634.

Gat, T., Liarzi, O., Skovorodnikova, Y., & Ezra, D. (2012). Characterization c Alternaria alternata causing black spot disease of pomegranate in Israel usin a molecular marker. Plant disease, 96(10), 1513-1518.

Jadhav, V. T., & Sharma, K. K. (2009, June). Integrated management of disease in pomegranate. In Souvenir and abstracts 2nd international symposium o pomegranate and minor including Mediterranean fruits, UAS Dharwad (pp. 23-27

Munhuweyi, K., Lennox, C. L., Meitz-Hopkins, J. C., Caleb, O. J., & Opara, U. I (2016). Major diseases of pomegranate (Punica granatum L.), their causes an management—A review. Scientia Horticulturae, 211, 126-139.

Özturk, N., & Ulusoy, M. R. (2006, October). Pests and natural enemies determined i pomegranate orchards in Turkey. In I International Symposium on Pomegranat and Minor Mediterranean Fruits 818 (pp. 277-284).

Shevale, B. S., & Kaulgud, S. N. (1998). Population dynamics of pests of pomegranat Punica granatum Linnaeus. In Advances in IPM for horticultural crop: Proceedings of the First National Symposium on Pest Management i Horticultural Crops: environmental implications and thrusts, Bangalor India, 15-17 October 1997. (pp. 47-51). Association for Advancement of Pes Management in Horticultural Ecosystems, Indian Institute of Horticultura Research.

21

Sapota

INSECT PESTS

Bud and fruit borer, *Nephopteryx eugraphella*

Damage symptoms

- » Most sapota pests don't bother with the fruits, but this one can ruin your buds, blooms, and foliage.

- » Caterpillars prefer leaves, although they can also be found on buds, flowers, and even vulnerable fruits.

- » The caterpillars weave a web of leaves and then consume the chlorophyll within, leaving behind a delicate web of veins.

- » They also eat into the insides of flower buds and young fruits, causing them to shrivel and die before the caterpillars move on to another target.

- » Clusters of dead leaves and flower buds hanging from webbed shoots are a telltale sign of this pest's invasion.

- » The pest's activity levels rise with the emergence of new growth and blossoms, but it can be found at any time of year.

Pest identification

- » **Egg** - Lightly coloured, oval, and the color and form of an egg.

- » **Larva** - reddish in color and have brown lines running lengthwise over their bodies.

- » **Adult** - Moth with a brown or black patch on its forewings as an adult

Management

- » Get rid of and burn all the infected masses.
- » Those contaminated apples and oranges need to be disposed of immediately
- » Gather the dried leaf web clusters and throw them away.
- » Apply a mixture of Phosalone 35 EC 2 ml/L, Phosphamidon 40 SL 2 ml/L Neem seed kernel extract 5%, and Chlorpyrifos 20 EC 2.5 ml/L.

Seed borer, *Trymalitis margarias*

Damage symptoms

- » The endosperm is its only source of nutrition.
- » Mature larvae excavate a passage out of the fruit in preparation for the pupation process.
- » Fungi and ants use this opening to infest and spoil produce.
- » There has been reported damage to ripe fruits at the rate of 40-80 %.
- » Because of this insect, fruit quality declines, causing a drop in market price and a loss of credibility for farmers.

Pest identification

- » **Larva:** Very tiny, whitish with a reddish tinge

Management

- » The marble to lime size of fruits was identified as the critical stage for IPM intervention.
- » It was found that a 0.028% concentration of decamethrin was the most effective medication for management.

Green scale, *Coccus viridis*

Damage symptoms

- » These scales are small, oval, and greenish-yellow in colour. They feed on the plant's leaves or young, fragile branches and shoots.
- » Black sooty mould appears on the leaves and twigs as a result of honeydew secreted by the insect, which invites subsequent fungal infection.
- » Summer is peak season for most infestations.

Pest identification

» **Eggs:** Eggs are deposited singly and are a whitish green colour.

» **Nymphs:** six-legged, flat, yellowish green creatures with an oval body shape.

» **Adults:** a glossy, light green colour with a prominent black, irregular U-shaped internal marking that is clearly apparent on the dorsal side. There are also two black specks, one on each of the lower eyelids, that are visible with a hand lens. Elongated oval with a gentle convexity characterises the general outline shape. The adult's scale length is between 2.5 and 3.25 mm.

Management

» While the infection is still in its non-fruiting stage, prune off afflicted areas and burn.

» In the event of a severe infestation, spray the plant during the dormant or early fruiting season with either Diazinon 20 EC 4 ml/L or Quinalphos 25 EC @ 2 ml/L after pruning the affected areas.

» When even a little infestation is detected, dusting the tree's base with 2% Methyl parathion @ 25g can stop the ants from climbing up the tree.

» To stop the transmission of pests to uninfested trees, cut down any branches that touch infected ones.

» In order to get rid of the sooty mould, spray it with 2% starch.

» January/February 10 grubs per tree release of *Cryptolaemus montrouzieri*.

» The aphelinid *Coccophagus* sp. had a crucial role in controlling the soft green scale population on sapota by acting as a parasite.

Bud worm, *Anarsia achrasella*

Crop losses

» The bud borer is known to cause significant damage, with estimates ranging from 2-15%.

Damage symptoms

» The caterpillar spins webs and dines on foliage.

» Tossing out chlorophyll counts.

» Dehydrated leaves dangling on stalks with webbed tips.

» Fruit production is diminished as a result of this pest's infestation of flower buds and other floral organs, which results in holes being chewed out of the ovary and the petals.

» Flora and buds with webbed petals.

» Destruction of flower buds and blossoms.

» Flowers that have been gnawed on by a bore have holes and faeces.

Pest identification

» **Egg:** Laid smooth and oval, and is initially white, but gradually turns a light brown colour before hatching.

» **Larva** - Tiny, slender, pinkish brown, with a black head

» **Adult:** a grey moth with a black spot on its wings.

Management

» Collecting and destroying damaged leaf webs containing larvae is necessary.

» Fenvalerate 20 EC 1 ml/L, Phosalone 35 EC 2 ml/L, or Phosphamidon 40 SL 2 ml/L can all be sprayed.

» The use of a spray containing 0.025% cypermethrin provides protection.

» Use a 5% extract of Neem seed kernels or 2% oil for spraying.

Fruit fly, *Bactrocera (Dacus) dorsalis*

Damage symptoms

» Maggots cause rot patches and fruit loss by eating through ripe fruit.

» Fluid seepage is observed.

» Spots of brownish decay on fruits.

Pest identification

» **Egg** - 0.8 mm in length, 0.2 mm in width, with a little protrusion of the micropyle at the front.

» **Larva** - these insects are apodous maggots that are a dull yellow in colour.

» **Adult** - Transparent brown wings and a light brown body

Management

» Collect the rotten fruit and bury it in a pit to get rid of the pests.

» Pupae might be exposed if ploughing is done throughout the summer.

» Fly activity can be tracked with methyl eugenol sex lure traps.

» The following concentrations of Fenthion, Malathion, Dimethoate, and Carbaryl should be sprayed: 100 EC 2 ml/ L, 50 EC 2 ml/ L, 1 ml/ L, and 4 g/ L, respectively.

» Dispersal of Predators and Prey *Opium*, in addition to the *Philippine Spalangia*, serves as a compensatory drug.

» Methyl eugenol 1% solution and Malathion 0.1% solution can be used to make bait. Put 10 cc of the mixture into each trap and distribute them throughout 25 locations in a single hectare.

» Apply Pyrethrum dust and sprays for extreme infestations.

Stem borer, *Plocaederus ferrugineus*

Damage symptoms

» Small pinpricks seen in the collar area.

» Gummosis.

» Frass escaping via the collar's bore holes.

» Turning yellow and dropping its leaves.

» The tree's twigs wither and eventually die.

Pest identification

» **Egg:** Ovoid, dirty white egg hidden beneath flaky bark.

» **Grub:** Size of a fully grown grub is about 7.5 cm.

» **Adult:** The adult beetle is a tanner's reddish brown colour. Dark brown to virtually black on the top and middle of the body.

Management

» Get rid of all the dead or dying vegetation.

» Implement proper sanitation measures in the field.

» Swab the lower trunk section with a mixture of coal tar and kerosene (1:2) or Carbaryl 50 WP (20 g/L) after removing loose bark (1m height).

» Cotton soaked with Monocrotophos 36 WSC at a rate of 10 ml per 2.5 cm per tree is used as padding.

» Remove grub with a hook, then fill each hole with mud, apply Monocrotophc (10–20 ml), a Celphos tablet (3 g Aluminum phosphide), and Carbofura 3G (5 g).

» Copper oxychloride paste should be applied to the tree trunk if infestation are severe.

» *Chrysoperla zastrowi sillemi*, Geocoris sp. big-eyed bugs, and pentatomi bugs are all potential predators *(Eocanthecona furcellata)*.

Striped mealy bug, *Ferrisia virgata*

Damage symptoms

» Underneath the leaf surface and on the tips of the shoots, a white meal material can be seen.

» Older leaves start to turn yellow.

Pest identification

» Adult - An adult of these wingless, soft-bodied beetles measures aroun 1/20 to 1/5 of an inch in length.

» A female is easily identified by her lengthy filaments at the tail end.

Management

» The egg masses and caterpillars must be collected and disposed of.

» Gathered larvae can be easily killed with a blazing torch.

» Catch and kill adults with a light trap.

» Mist the area with either 0.05% Methyl demeton 25 EC or 0.06% Dimethoat 30 EC.

» Twenty *Cryptoleamus montrouzieri* plants will be planted each tree in the wild

Hairy caterpillar, *Metanastria hyrtaca*

Damage symptoms

» Caterpillars gorge themselves on foliage.

» Foliage loss in trees.

Pest identification

» Larvae - have a drab brown colour with black patches and protruding tuft of hair along the sides.

» Adults: Head and torso are grey, but the abdomen is white in adults. Its reddish brown, spotty, white forewings feature white edging. White tails and wings.

Management

» Implement proper sanitation measures in the field.

» The trees should be kept clean of weeds and other obstacles.

» Get rid of the egg clumps by collecting them.

» Torching clusters of tree-dwelling larvae and discarding the ashes.

» Spraying with a 0.025% Cypermethrin solution.

» We recommend spraying at a rate of 2 ml/L of either Chlorpyriphos 20EC, Endosulfan 35EC, or Phosalone.

» Carbaryl 10 D being dusted onto the main stem and branches (4 feet around the tree).

» Predators include pentatomid bugs and big-eyed bugs (*Geocoris* sp) (*Eocanthecona furcellata*).

» *Brachymeria* sp., a chalcidid wasp, was released into the wild.

DISEASES

Leaf spot, *Phaeophleospora indica*

Symptoms

» At maturity, the pathogen causes a proliferation of tiny, circular, pink to reddish brown spots on the leaf, all of which have white centres.

» Spots join forces; infected leaves become yellow and fall off too soon.

» Defoliation has also been observed to cause a significant decrease in yield.

Favorable conditions

» Diseases tend to flourish in wetter climates.

Survival and spread

» The inoculum found in the seeds is what first causes infection, as the pathogen is a fungus. Fungi can also make do with glumes, fruit, and other bits of plant waste.

Management

» Mulching defoliated leaves and then applying fungicide spray is more effective than either method used separately.

» Mancozeb 0.25 % or copper oxychloride 2 % spray.

» It was shown that monthly spraying with 0.2% Dithane Z-78 or 0.3% Lonacol followed by 0.5% Blitox and 0.2% Cuman L was effective.

» In new plantations, you should avoid growing vulnerable sapota cvs. like Cricket Ball and Kinhabarthi.

Sooty mold, *Capnodium* sp.

Symptoms

» Aphid and scale insect excrement, which resembles honeydew, is the source of this fungal disease.

» The fungus spreads slowly across the leaf, eventually interfering with photosynthesis to a devastating degree.

» Because of this, fewer nutrients are transported to the fruits, reducing their size.

Favorable conditions

» The conditions for the spread of disease are optimal in a humid and damp environment.

Survival and spread

» The honey dew secretions of the scale insects provide the ideal medium for the fungus to thrive, and so determine the degree of infection.

» Ascospores disperse through the air and cause the disease.

Management

» Spray a starch solution of 5% on any mould or mildew spots.

» Use systemic pesticide sprays to get rid of the pests.

» Spraying with a solution of 40 gm of zineb to every eighteen litres of water is quite effective.

Fasciation/Flat limb, *Botryodiplodia theobromae*

Symptoms

- » Damaged trees' limbs become misshapen and twisted, with visible scars.
- » Reduced in size and turning yellow, the leaves fall off.
- » Leaves and blossoms cluster on afflicted branches.
- » Premature loss of foliage and fruits, as well as the continued smallness and dryness of the fruits themselves, is a common occurrence.
- » The flowers have not yet produced any fruit.
- » When fruit does set, it is little and unripe.

Management

- » The disease can be contained by removing and killing the distorted shoots.
- » Disease rates can be lowered by spraying with either 0.1% Carbendazim or 0.2% Mancozeb.
- » Leaves and fruit should be sprayed once every two weeks.

References

Bisane, K. D. (2019). Seasonal variability of chiku moth, Nephopteryx eugraphella (Ragonot) in relation to ecological parameters and crop phenology of sapota. Pest Management in Horticultural Ecosystems, (1), 37-43.

Mani, M., & Jayanthi, P. K. (2022). Pests and Their Management on Sapota/Sapodilla. Trends in Horticultural Entomology, 655-670.

Patil, B. J., Suchithra Kumari, M. H., Revannavar, R., Shivaprasad, M., HS, Y. K., & Hanumantharaya, L. (2020). Seasonal incidence of sapota seed borer, Trymalitis margarias Meyrick (Lepidoptera: Tortricidae) in Mudigere. Journal of Entomology and Zoology studies, 8(6): 1267-1274.

Patil, P. D., Dhane, A. S., Karande, R. A., Raut, R. A. and Dalvi, M. B. (2019). Integrated management of fruit drop in sapota (Manilkara zapota). IJCS, 7(3), 4249-4252.

Pattanayak, S., & Das, S. A perspective on Integrated Disease Management Strategies in Minor Tropical Fruit Crops of India.

Somwanshi, S. D., Magar, S. J. and Suryawanshi, A. P. (2021). Postharvest Diseases of Minor Fruits and Their Management. In Postharvest Handling and Diseases of Horticultural Produce (pp. 295-304). 6000 Broken Sound Parkway NW, Suite 300, Boca Raton, FL 33487-2742: CRC Press.

22

Strawberry

INSECT PESTS

Cut worms, *Agrostis ipsilon*

Damage symptoms

- » It is nocturnal and feeds on the leaves and stems of strawberry bushes, typically near the ground.
- » Holes of varying sizes and shapes in the leaves are caused by leaf feeding, however this has little to no effect on crop yields.
- » Significant damage can be caused by larvae severing young plants; in fact, a single larva might sever many plants in a single night.
- » Strawberries' crown size and productivity can be drastically cut by stem feeding.
- » Older plants will suffer more damage.

Pest identification

- » The adults of these moths are around 1.5 inches in length and brown or grey in coloration.
- » Larvae that have reached maturity are about 1.5 inches in length, are sturdy, and have smooth skin that is a mottled brown or grey.
- » When disturbed, larvae often drop to the floor and coil into a C shape.

Management

- » Eliminating weeds is crucial in stopping a cutworm infestation.
- » Large numbers of male cutworms can be captured using pheromone traps in the late summer and early fall. These traps consist of a 5- by 7-mm rubber

septum charged with 30 μm of (Z)-7-dodecen-1-y1 acetate and 10 μm of (Z)-9- tetradecen-1-yl acetate.

» Apply *Bacillus thuringiensis* or Spinosad Entrust.

Armyworm, *Spodoptera exigua*

Damage symptoms

» Make holes in the foliage that can range in shape from round to erratic.

» Heavy damage to the leaf tissue caused by immature larvae.

» Create small, dry punctures on the fruit.

» Strawberries planted in the summer or fall suffer the most damage from beet armyworm.

» Beet armyworm larvae skeletonize the top or lower leaf surfaces next to their egg mass in order to eat.

» Larvae feed on plant matter before moving on to fruit.

» Young plants' tops are easy targets for larger larvae.

Pest identification

» While young larvae are a light green to yellow, older larvae are a darker green with a dark and light line running along the side of their body and a pink or yellow underbelly.

Management

» Because adult moths seek out weeds in which to deposit their eggs, reducing weed growth in and around the field can reduce armyworm populations.

» Lorsban has a moderate impact on beneficial insect and mite populations while providing fair to good control of worm pests in strawberries.

» Among worm treatments, Lannate is among the best in eradicating cutworms and beet armyworms, though Brigade and Danitol are also effective.

» Biological control using natural enemies that prey on armyworm larvae and the application of *Bacillus thuringiensis* are two examples of organic techniques of controlling armyworms.

» The ichneumonid parasite *Hyposoter exiguae* provides biological control, and a virus that frequently infects armyworms can result in significant mortality rates, making it necessary to eradicate the pest.

Weevils, *Otiorhynchus* spp.

Damage symptoms

- » Adults feed on the tips of leaves while the larvae eat the roots.
- » Unless it serves as a warning sign of an infestation, damage caused by adult weevils while they snack on leaves and fruits is usually of little consequence
- » When larvae feed on roots, they cause the most damage.
- » Fruit is small and seedy, and plants are stunted.
- » Drought conditions in the summer can be disastrous for damaged plants.
- » Infestations of this magnitude can wipe out an entire strawberry crop in single growing season.

Pest identification

- » The adult stage of this insect is a snot-nosed beetle, and its coloration range from dark to light brown.
- » Underground root-feeding grubs of a creamy white tint are the larvae.

Survival

- » The dirt is a safe place for the larvae to spend the winter.

Management

- » Eliminate any grass or weeds that may grow in strawberry beds.
- » In order to reduce the likelihood of attracting weevils, you should not plant near wooded areas or bushes that bear either blackberries or elderberries.
- » Strawberry weevils can be effectively controlled by using pesticide spray or dusts containing Pyrethroids.

Strawberry aphid, *Chaetosiphon fragaefolii*

Damage symptoms

- » Leaves may turn yellow, have deformed necrotic patches, and/or have their growth halted due to a heavy aphid infestation.
- » Aphids produce honeydew, a sugary secretion that promotes the growth of sooty mould on plants.
- » Aphids are a big problem because they spread a number of diseases that can severely impact strawberry yields.

Management

» If an aphid infestation is contained to a small area, such a few leaves or shoots, it can be easily removed through pruning.

» It's important to inspect transplants for aphids before planting them.

» Aphids can be discouraged from feeding on plants by using reflective mulches, such as silver coloured plastic.

» A powerful stream of water can be used to blast aphids off the leaves of hardy plants.

» The most effective means of prevention are insecticidal soaps and oils like Neem and Canola.

Whitefly, *Trialeurodes vaporariorum*

Damage symptoms

» While feeding on plant juices, whiteflies emit honeydew, which can support the growth of sooty mould fungus in dense populations.

» Sometimes populations grow to unsustainable levels, leading to monetary losses.

Management

» In locations where whiteflies are a major concern, plantings that span more than one year should be avoided.

» New strawberry fields should be planted when the old hosts have been eradicated.

» Although useful, trap crops have shown only moderate success, and in some cases have even been used as bridge hosts.

» Sticky tape is used to keep track of the pests.

» Have fun releasing parasitic wasps into the strawberry crops.

» Whitefly control with Admire is adequate.

» Despite being a contact insecticide, lannate does a decent job at killing adult whiteflies.

» Only moderate adult whitefly control can be expected when using Danitol in conjunction with Malathion.

Spider mite, *Tetranychus urticae*

Damage symptoms

> » Spotting of yellow on the foliage; the leaves may seem bronzed; a layer of webbing

> » A mite's web or the underside of a leaf may be crawling with tiny moving dots

> » The plant's leaves may become yellow and fall off.

Favorable conditions

> » Dusty environments are perfect for spider mites, and plants that are already struggling to survive due to a lack of water are easy prey.

Management

> » One way to prevent spider mites from taking over a plant is to spray it with a powerful stream of water.

> » Use insecticidal soap on plants if pests like mites become an issue.

> » Horticultural oil is effective against all stages of mites, including eggs, and kills by suffocation.

> » Dicofol (Kelthane), abamectin (AgriMek), bifenazate (Acramite), venbutatin-oxide (Vendex), and hexythiazox are all examples of miticides (Savey).

> » Pesticide that uses living organisms as its vector If *Phytoseiulus persimilis* is introduced to your garden early enough, you won't have to worry about spider mites. The early part of the season is when this natural opponent really shines as a specialist. *Neoseiulus californicus* is a generalist predator that thrives in later seasons because it is able to adapt to a wide range of temperatures and humidity levels.

DISEASES

Anthracnose, *Colletotrichum fragariae*

Symptoms

> » On the petioles and the runner stems, dark, elongated sores appear.

> » Infected crown tissue can cause a condition known as crown rot, which can ultimately result in the death of the entire plant.

> » Lesions on fruit initially appear as water-soaked white, tan, or light brown spots up to 3 mm in diameter.

» Within two to three days, the lesions turn brown or dark brown, become sunken, and spread across the majority of the fruit.

Survival and spread

» Soil fungi can live for up to 9 months underground.

» Splashing water and soil onto newly planted plants spreads the disease.

Management

» The inoculum in the soil could be lost if it is solarized.

» Plants should be disease-free.

» Substituting non-host crops in a rotation.

» Water splashes serve as a vector for the dissemination of fungi. Drip irrigation is preferable to overhead sprinkler systems.

» It is recommended to use straw as mulch in perennial matted row plantings to lessen the amount of water splashed around and the likelihood of disease transmission.

» Get rid of all the rotting, disease-ridden plant matter you can.

» Fungicides are most effective when used before disease becomes a problem.

Leaf spot, *Mycosphaerella fragariae*

Symptoms

» Lesions appear on the upper and lower leaf surfaces and can be either circular or irregular in shape, with a deep purple border and a gray-white core.

» In sensitive kinds, lesions can spread and become rather large, yet the lesion's centre always remains brown.

» Fruits, petioles, and stolons are all susceptible to lesion development.

Spread

» Disease was disseminated via water splashes.

Management

» Produce plants using virus-free seedlings.

» Use a foliar fungicide for protection.

Red root rot/Red stele, *Phytophthora fragariae*

Symptoms

» The plant growth rates will slow down, and they'll take on a dreary, bluish-green color.

» Plants will recover from their wounds slightly come spring.

» Wrinkled and possibly tinted red, yellow, or orange, old leaves have fallen off.

» Miniature in size, new leaves are.

» No or very few flowers will form on a diseased plant.

» The little fruit will dry up and shrivel.

» There is a deficiency of root hairs.

» A reddening of the central cylinder can be seen when cutting into the primary roots.

» The root core's reddish color may spread to the crown.

Survival and spread

» Soil fungi can live for up to 9 months underground.

» Splashing water and soil onto newly planted plants spreads the disease.

Management

» The inoculum in the soil could be lost if it is solarized.

» Plants should be disease-free.

» Substituting non-host crops in a rotation.

» Water splashes serve as a vector for the dissemination of fungi. Drip irrigation is preferable to overhead sprinkler systems.

» It is recommended to use straw as mulch in perennial matted row plantings to lessen the amount of water splashed around and the likelihood of disease transmission.

» Get rid of all the rotting, disease-ridden plant matter you can.

» Fungicides are most effective when used before disease becomes a problem.

Root rot, *Rhizoctonia fragariae*

Symptoms

- » Overall weakness, manifested by stunted runner development and little fruit.

- » In times of high-water demand, such as when a plant is actively growing, has just finished fruiting, or is under stress from a lack of water, such as during a drought, plants might die.

- » The tips of the roots may decay, and the white roots may be speckled with black blemishes.

- » In its early phases, the root core is white rather than red (which indicates red stele disease).

- » Both the inner and outer layers of tissue will turn black if plants are seriously harmed.

Management

- » Two-month solarization treatment for the soil.

- » Cover crops (ryegrass, Sudan grass, and sorghum/Sudan) used in a two-year crop rotation.

- » Choose only plants that have white roots and are in good health.

- » Ammonium-based nitrogen fertilisers reduced the incidence of black root rot.

- » Success has been seen with *Trichoderma harzianum* (Root Shield).

- » Applying a biocide to the soil, like methyl bromide or chloropicrin, that has a wide range of activity.

Grey mold, *Botrytis cinereria*

Symptoms

- » Flower petals, fruit, petioles, leaf petioles, and stems are all fair game for the fungi.

- » Rapid decomposition of infected flowers and fruit stalks is observed during the blossoming stage. This brown rot affects both unripe and ripe fruit.

- » Dry, greyish spores begin to coat the entire fruit as this progresses.

- » Fruit rot typically begins at the calyx end or on the edges of fruit that have come into contact with other rotting fruit, though it can begin anywhere on the fruit.

Favorable conditions

» Wet and cold with a low temperature and significant humidity.

Survival and spread

» Fungi overwinter on plant matter, infect flower parts, and then either destroy the fruit or dormantly infect the fruit until it reaches a ripe stage. At any point during the fruiting season, the spores that have been produced will germinate and infect nearby plants.

» Natural methods of dispersal include wind and water droplets from rain or sprinklers.

Management

» Straw mulch, which can be easily cleaned, helps keep berries off the dirt and thus cleaner, and also helps avoid fruit rot.

» The spread of this disease can be slowed by removing any berries that have reached full ripeness or have been affected.

» Fruits are cold stored at temperatures between 0 and 10 °C.

» Dead or dying plants should be promptly removed.

» Don't just toss out any old plants that have died.

» Humidity should be lowered by increasing airflow and decreasing plant crowding.

» Mulch using plastic to lessen the amount of soil that gets on your fruit.

» After harvesting, plough crop residue back into the soil.

» Vinclozolin and Benomyl can be used right up until harvest time.

Blue mold, *Pencillium expansum*

Symptoms

» Faint green sporulation of a fungus in a depressed part of the apple that looks squishy and mushy.

» Eventually, decayed tissue will be removed, leaving a hollow on the strawberry's side.

Management

» Biofungicides such as Aspire (yeast- *Candida oleophila*) and Biosave 110 are sprayed on fruit after harvest to prevent mould *(Pseudomonas syringae)*

Angular leaf spot, *Xanthomonas fragariae*

Symptoms

» Initially appearing as tiny water-soaked lesions on the lower surfaces of leaves, these sores eventually grow into dark green or translucent angular patches that spew germs.

» Sometimes, clusters of lesions will join together to form bright red areas surrounded by a chlorotic halo.

» Calyxes are not immune to the pathogen's infection.

Survival and spread

» Bacteria may overwinter on plant waste for lengthy periods of time but not in soil, thus it must be present in crop residue or on plants that are kept indoors.

» Splashing water can disperse bacteria.

Management

» Only use verified seedlings in your plantings.

» Replace host plants with non-host plants in your crop rotation.

» The use of overhead sprinklers should be avoided.

Fire blight, *Erwinia amylovora*

Symptoms

» Infected leaves dry out and turn brown or black, yet they don't fall off the trees.

» Oftentimes, vegetative sprouts will wither and die.

» As the disease worsens, the colour of infected flowers may shift from red to brown to black.

» Fruit that has been damaged and is subsequently infected usually shows signs of infection in the form of red, brown, or black blemishes.

Management

» Any cankers that have survived the winter should be cut out now, when the plant is dormant.

» Pruning should begin at least 6-8 inches below the discoloured area to remove active lesions.

» To prevent the spread of bacteria, pruners should be disinfected after each cut

» Spread Agrimycin, Copper oxychloride, or Dithane M-45 on the ground.

References

Alford, D. V. (1976). Some observations on the effect of pests on strawberry yields. *Annals of Applied Biology*, *84*(3), 440-444.

Easterbrook, M. A., Fitzgerald, J. D., Pinch, C., Tooley, J., & Xu, X. M. (2003). Development times and fecundity of three important arthropod pests on strawberry in the United Kingdom. *Annals of Applied Biology*, *143*(3), 325-331.

Lahiri, S., Smith, H. A., Gireesh, M., Kaur, G., & Montemayor, J. D. (2022). Arthropod pest management in strawberry. *Insects*, *13*(5), 475.

Paulus, A. O. (1990). Fungal diseases of strawberry. *HortScience*, *25*(8), 885-889.

Solomon, M. G., Jay, C. N., Innocenzi, P. J., Fitzgerald, J. D., Crook, D., Crook, A. M. ... & Cross, J. V. (2001). Natural enemies and biocontrol of pests of strawberry in northern and central Europe. *Biocontrol Science and Technology*, *11*(2), 165-216.

Strand, L. L. (2008). *Integrated pest management for strawberries* (Vol. 3351). UCANR Publications.

Subbarao, K. V., Kabir, Z., Martin, F. N., & Koike, S. T. (2007). Management of soilborne diseases in strawberry using vegetable rotations. *Plant Disease*, *91*(8) 964-972.

Tzanetakis, I. E., Halgren, A. B., Keller, K. E., Hokanson, S. C., Maas, J. L., McCarthy P. L., & Martin, R. R. (2004). Identification and detection of a virus associated with strawberry pallidosis disease. *Plant Disease*, *88*(4), 383-390.

Vaughan, E. K. (1933). Transmission of the Crinkle Disease of Strawberry. *Phytopathology*, *23*(9).

Weber, R. W., & Hahn, M. (2019). Grey mould disease of strawberry in northern Germany: causal agents, fungicide resistance and management strategies. *Applied microbiology and biotechnology*, *103*(4), 1589-1597.

For Product Safety Concerns and Information please contact our EU
representative GPSR@taylorandfrancis.com
Taylor & Francis Verlag GmbH, Kaufingerstraße 24, 80331 München, Germany

www.ingramcontent.com/pod-product-compliance
Ingram Content Group UK Ltd.
Pitfield, Milton Keynes, MK11 3LW, UK
UKHW021123180425
457613UK00006B/206